OPPORTUNITIES IN THE HYDROLOGIC SCIENCES

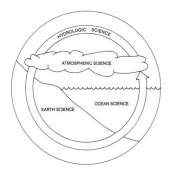

Committee on Opportunities in the Hydrologic Sciences
Water Science and Technology Board
Commission on Geosciences, Environment, and Resources
National Research Council

NATIONAL ACADEMY PRESS
Washington, D.C.   1991

NATIONAL ACADEMY PRESS   2101 Constitution Avenue, NW   Washington, DC 20418

NOTICE: The project that is the subject of this report was approved by the Governing Board of the National Research Council, whose members are drawn from the councils of the National Academy of Sciences, the National Academy of Engineering, and the Institute of Medicine. The members of the committee responsible for the report were chosen for their special competences and with regard for appropriate balance.

This report has been reviewed by a group other than the authors, according to procedures approved by a Report Review Committee consisting of members of the National Academy of Sciences, the National Academy of Engineering, and the Institute of Medicine.

The National Academy of Sciences is a private, nonprofit self-perpetuating society of distinguished scholars engaged in scientific and engineering research, dedicated to the furtherance of science and technology and to their use for the general welfare. Upon the authority of the charter granted to it by the Congress in 1863, the Academy has a mandate that requires it to advise the federal government on scientific and technical matters. Dr. Frank Press is president of the National Academy of Sciences.

The National Academy of Engineering was established in 1964, under the charter of the National Academy of Sciences, as a parallel organization of outstanding engineers. It is autonomous in its administration and in the selection of its members, sharing with the National Academy of Sciences the responsibility for advising the federal government. The National Academy of Engineering also sponsors engineering programs aimed at meeting national needs, encourages education and research, and recognizes the superior achievements of engineers. Dr. Robert M. White is president of the National Academy of Engineering.

The Institute of Medicine was established in 1970 by the National Academy of Sciences to secure the services of eminent members of appropriate professions in the examination of policy matters pertaining to the health of the public. The Institute acts under the responsibility given to the National Academy of Sciences by its congressional charter to be an adviser to the federal government and, upon its own initiative, to identify issues of medical care, research, and education. Dr. Samuel O. Thier is president of the Institute of Medicine.

The National Research Council was organized by the National Academy of Sciences in 1916 to associate the broad community of science and technology with the Academy's purpose of furthering knowledge and advising the federal government. Functioning in accordance with general policies determined by the Academy, the Council has become the principal operating agency of both the National Academy of Sciences and the National Academy of Engineering in providing services to the government, the public, and the scientific and engineering communities. The Council is administered jointly by both Academies and the Institute of Medicine. Dr. Frank Press and Dr. Robert M. White are chairman and vice chairman, respectively, of the National Research Council.

Support for this project was provided by the National Research Council, the U.S. Geological Survey and the National Weather Service under Contract No. 14-08-0001-G1506, the National Science Foundation under Grant No. EAR-8719003, the National Aeronautics and Space Administration and the Army Research Office under Contract No. NAGW-1310, the U.S. Forest Service under Agreement No. 90-G-011/R, and The Mobil Corporation.

Library of Congress Cataloging-in-Publication Data

Opportunities in the hydrologic sciences / Committee on Opportunities
    in the Hydrologic Sciences, Water Science and Technology Board,
    Commission on Geosciences, Environment, and Resources, National
    Research Council.
        p. cm.
    Includes bibliographical references and index.
    ISBN 0-309-04244-5
    1. Hydrology—Vocational guidance. I. National Research Council
(U.S.). Committee on Opportunities in the Hydrologic Sciences.
GB665.0315 1991
551.46'0023—dc20   90-49577
                                                          CIP

Cover art reproduced with permission from Sally J. Bensusen. Copyright © 1990 by Sally J. Bensusen.

Committee on Opportunities in the Hydrologic Sciences

PETER S. EAGLESON, Massachusetts Institute of Technology, *Chairman*
WILFRIED H. BRUTSAERT, Cornell University
SAMUEL C. COLBECK, U.S. Army Cold Regions Research and
  Engineering Laboratory, Hanover, New Hampshire
KENNETH W. CUMMINS, University of Pittsburgh
JEFF DOZIER, University of California-Santa Barbara
THOMAS DUNNE, University of Washington
JOHN M. EDMOND, Massachusetts Institute of Technology
VIJAY K. GUPTA, University of Colorado-Boulder
GORDON C. JACOBY, Lamont-Doherty Geological Observatory,
  Palisades, New York
SYUKURO MANABE, National Oceanic and Atmospheric
  Administration, Princeton, New Jersey
SHARON E. NICHOLSON, Florida State University
DONALD R. NIELSEN, University of California-Davis
IGNACIO RODRIGUEZ-ITURBE, University of Iowa
JACOB RUBIN, U.S. Geological Survey, Menlo Park, California
J. LESLIE SMITH, University of British Columbia
GARRISON SPOSITO, University of California-Berkeley
WAYNE T. SWANK, U.S. Department of Agriculture, Coweeta
  Hydrologic Laboratory, Otto, North Carolina
EDWARD J. ZIPSER, Texas A & M University

Ex-Officio

STEPHEN BURGES, University of Washington (WSTB member
  through June 1989)

National Research Council Staff

STEPHEN D. PARKER, Project Manager
WENDY L. MELGIN, Staff Officer (through October 1989)
RENEE A. HAWKINS, Project Secretary
SUSAN MAURIZI, Editor

Liaison Representatives

GHASSEM ASRAR, National Aeronautics and Space Administration,
  Washington, D.C.
JOHN A. MACCINI, National Science Foundation, Washington, D.C.
STEVEN MOCK, U.S. Army Research Office, Research Triangle Park,
  North Carolina
MARSHALL MOSS, U.S. Geological Survey, Tucson, Arizona
JOHN SCHAAKE, National Weather Service, Silver Spring, Maryland

Foreword

In 1982, soon after the beginning of my term as president of the National Academy of Sciences (NAS) and chairman of the National Research Council (NRC), a major reorganization of the NRC's structure was implemented. As part of this reorganization, we created the Water Science and Technology Board (WSTB) to recognize the importance of water resources to the nation.

The driving forces behind the WSTB's establishment were threefold: a unit of the NRC specifically assigned to water resources would emphasize their national importance; the complexity of water science and technology issues lends itself well to the NRC's ability to approach problems in an interdisciplinary manner; and, perhaps most important, the field needed sounder scientific underpinnings, particularly as we begin to take a more global and system-oriented view of our environment.

Over the past several years there has been increasing concern among scientific hydrologists about the future and long-term vitality of their field. This is owing, somewhat paradoxically, to the fact that throughout the history of this field applications have preceded science. Civil and agricultural engineers are in large part responsible for the high level of water-related health and safety enjoyed by modern urban societies of the developed world. Nevertheless, this pragmatic focus has left fundamental hydrologic science lagging behind in comparison with other geosciences. The result is a scientific and educational base in hydrology that is incompatible with the scope and complexity of many current and emerging problems.

Many currently important surface hydrologic problems are so large in scale that the land surface and atmosphere must be treated as an interactively coupled system. Examples are the environmental impacts of tropical deforestation, large-scale irrigation and drainage, and acid precipitation. Prospects of climate change require forecasting based on global-scale understanding and have heightened interest in ancient hydroclimatology as revealed through paleohydrology.

Contemporary ground water problems are often large scale—from one to hundreds of kilometers. They involve major geological heterogeneity and complex issues of water chemistry. Examples are the containment and reduction of pollution, underground storage of toxic waste, aquifer recharge, geothermal power production, and the conjunctive management of surface and ground water systems both at local and regional scales.

The interdisciplinary nature of these problems requires increased application of principles from the atmospheric, geologic, chemical, and biological sciences; their geoscience perspective reveals important deficiencies in our basic knowledge of hydrologic science. Questions of scaling, equilibrium, stability, teleconnections, and space-time variability demand a renewed emphasis on fundamental hydrologic research. The needed understanding will be built from coordinated, long-term data sets (both at fine and large spatial scales) and founded on an educational base in the geosciences.

This report should be an important reference work on opportunities in the hydrologic sciences. It is intended to help guide science and educational policy decisions and to provide a scientific framework and research agenda for scientists, educators, and students making career plans. We hope it will also be of interest to the informed lay public. The document transmits the importance of the hydrologic sciences and identifies needed improvements to the research and educational infrastructure. If its recommendations are followed, we believe the strengthened scientific base of hydrology will contribute directly to improved management of water and the environment.

FRANK PRESS, *Chairman*
National Research Council

Preface

Hydrologic science deals with the occurrence, distribution, circulation, and properties of water on the earth. It is clearly a multidisciplinary science, as water is important to and affected by physical, chemical, and biological processes within all the compartments of the earth system: the atmosphere, glaciers and ice sheets, solid earth, rivers, lakes, and oceans. Because of this geophysical ubiquity, concern for issues of hydrologic science has been distributed among the traditional geoscience disciplines. As a result, an infrastructure of hydrologic science (i.e., a distinct discipline with a clear identity and supporting educational programs, research grant programs, and research institutions) has not developed, and a coherent understanding of water's role in the planetary-scale behavior of the earth system is missing. This report describes this problem and offers a set of recommended remedial actions.

THE PROBLEM

Water moves through the earth system in an endless cycle that forms the framework of hydrologic science. In so moving, it plays a central role in many of the physical, chemical, and biological processes regulating the earth system, where human activity is now inseparable from natural events. Water vapor is the working fluid of the atmospheric heat engine: through evaporation and condensation it drives important atmospheric and oceanic circulations and redistributes absorbed solar energy. As the primary greenhouse gas it is instrumental in setting planetary temperature. Through fluvial erosion and sedimentation, water, together with tectonics, shapes the land surface. Water is the universal solvent and the medium in which most changes of matter take place; hence it is the agent of element cycling. Finally, water is necessary for life.

Investments in water resources management over the last century have helped provide the remarkable levels of public health and safety enjoyed by the urban populations of the developed world. Although we have spent lavishly to cope with the scarcities and excesses of water and to ensure its potability, we have invested relatively little in the basic science underlying water's other roles in the planetary mechanisms. This science, hydrologic science, has a natural place as a geoscience alongside the atmospheric, ocean, and solid earth sciences; yet in the modern scientific establishment this niche is vacant.

Because of the pervasive role of water in human affairs, the development of hydrologic science has followed rather than led the applications—primarily water supply and hazard reduction—under the leadership of civil and agricultural engineers. The elaboration of the field, the education of its practitioners, and the creation of its research culture have therefore been driven by narrowly focused issues of engineering hydrology. The scale of understanding has been modest—generally limited to surface drainage basins with areas of 10,000 km^2 or less. The committee's perception of the intellectual relationships among these water-relevant disciplines is presented in Figure 1.

Hydrology has not been cultivated as a geoscience because until now there has been no practical need to build a comprehensive understanding of the global water cycle. Moreover, the patches of scientific knowledge that support traditional small-scale engineering applications do not merge into the coherent whole needed to understand the geophysical and biogeochemical functioning of water at the regional and continental scales of many emerging problems. These problems include the possible geographical redistribution of water resources due to climate change, the ecological consequences of large-scale water transfers, widespread mining of fossil ground water, the effect of land use changes on the regional hydrologic cycle, the effect of non-point sources of pollution on the quality of surface and ground water at regional scale, and the possibility of changing regimes of regional floods and droughts.

Furthermore, the training of hydrologic scientists cannot be accomplished efficiently in educational programs dominated either by applications-oriented constraints or by undergraduate preparation in which engineering predominates. A thorough background in mathematics, physics, chemistry, biology, and the geosciences is neces-

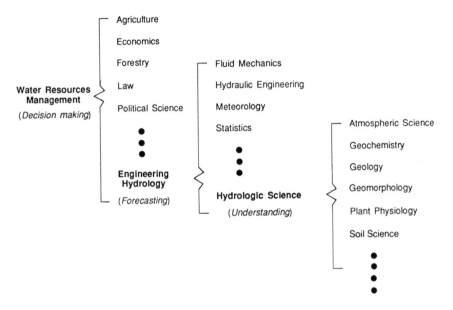

FIGURE 1 Water—the intellectual ingredients for its understanding, forecasting, and management.

sary. New institutional arrangements will be needed to allow student and faculty involvement in relevant field observations and to provide prompt access to the resulting data.

A COURSE OF ACTION

One step taken to help build the needed infrastructure for hydrologic science was the creation of the Committee on Opportunities in the Hydrologic Sciences (COHS) by the Water Science and Technology Board in January 1988. This committee was asked to conduct an assessment of the hydrologic sciences, including their definition, their current state of development, and their relationships with related geosciences and biosciences. The committee also was asked to identify promising new frontiers and applications and to outline an appropriate framework for education and research in the hydrologic sciences.

At the outset we should understand the committee's use of the term "geoscience" vis-à-vis "earth science" and the more recently coined "earth system science." The COHS follows the National Science Foundation (NSF) and uses "geoscience" to include atmospheric science, ocean science, solid earth science, glaciology, and, as argued herein, hydrologic science. As does the NSF, the committee interprets "earth science" as the solid earth sciences, including geology, petrology, seismology, volcanology, and so on. "Earth system science" includes all sciences relevant to the functioning of the planet earth as a set of interacting physical, chemical, and biological mechanisms. It differs from geoscience in its inclusion of important terrestrial biota and certain solar and other space physics effects on the earth, and in its emphasis on integrated planetary behavior.

To establish an identity for hydrologic science as a separate geoscience, the COHS defined its scope to include (1) the physical and chemical processes in the cycling of continental water at all scales as well as those biological processes that significantly interact with the hydrologic cycle, and (2) the spatial and temporal characteristics of the global water balance in all compartments of the earth system.

In presenting its findings the COHS has written for scientifically literate readers who are not necessarily hydrologists. We have avoided mathematics and lengthy scientific detail for the sake of clarity. The report also contains additional brief discussions of important practical problems whose solution will benefit from the anticipated scientific advances. In addition, to accent the human dimension of all scientific achievement, the committee has included throughout the volume short biographical vignettes of important past figures in hydrology. In summary, the report's contents are as follows:

Chapter 1, "Water and Life," explains the uniqueness and historical importance of water. It contains examples of the roles of water in civilization both as sustainer and hazard, and as a resource to be managed.

Chapter 2, "The Hydrologic Sciences," describes the evolution of the perception and definition of the hydrologic sciences and identifies as primary agents of change the increasing scale of applied problems and the concurrent spreading of anthropogenic influences. The hydrologic cycle is recognized to be the framework of these sciences, and its physical and chemical processes are illustrated. The status of understanding of these processes and of the biological components of the hydrologic cycle is summarized. Some major research questions are posed.

Chapter 3, "Some Critical and Emerging Areas," is the intellectual core of the report and contains a collection of essays on promising frontier research topics. In selecting these topics the committee has opted for the interesting and exciting, subjectively seeking to transmit the flavor of the science rather than to provide an exhaustive catalog of opportunities. A typical essay begins with a research question, is followed by a brief historical review, and concludes with a description of the problem and its importance to the science.

The essays are grouped in sets representing the major subdivisions of the hydrologic sciences. These subdivisions reflect the major theme of this report, namely that hydrologic science is a geoscience. Accordingly, the COHS looks at "Hydrology and the Earth's Crust," "Hydrology and Landforms," "Hydrology and Climatic Processes," "Hydrology and Weather Processes," "Hydrology and Surficial Processes," "Hydrology and Living Communities," "Hydrology and Chemical Processes," and "Hydrology and Applied Mathematics." There is a degree of deliberate redundancy and inconsistency in this arrangement. The quality of water is as much a part of the hydrologic cycle as is its mass or flow rate, and so chemical issues occur along with the physical in each of the geophysical subdivisions; indeed, several have been selected for presentation there. The same can be said for the mathematical topics of the final section, and, to a more limited degree, for biology. However, the committee has chosen to concentrate its discussions on the relation of hydrology to its sibling sciences in separate sections in order to call the attention of biologists, chemists, and mathematicians to interesting hydrologic problems.

Chapter 4, "Scientific Issues of Data Collection, Distribution, and Analysis," discusses the need for, the characteristics of, and the current status of hydrologic data. It concludes with a set of brief essays concerning topics such as new technology, methods of analysis, and coordinated multidisciplinary experiments.

Chapter 5, "Education in the Hydrologic Sciences," contrasts education for the internally driven puzzle solving of science with that for the externally driven problem solving of engineering. It makes specific recommendations relative to hydrologic science for programs at the graduate, undergraduate, and kindergarten through twelfth grade levels.

Chapter 6, "Scientific Priorities," outlines a rational process for setting scientific priorities and presents a set of research, data, and educational opportunities that the committee believes are most important for hydrologic science at this time.

Chapter 7, "Resources and Strategies," closes the report with programmatic investment priorities for funding agencies, and with strategic actions that individual scientists and their scientific societies can take to enhance the image and status of hydrologic science.

The report concludes with four appendixes. Appendix A gives an estimate of recent annual investments of U.S. federal agencies in research in hydrologic science. Appendix B profiles the hydrologic science community, contrasting the results of a 1988 human resources questionnaire with those of a similar survey published in 1962. Appendix C acknowledges the many contributors to this report in addition to those listed in the front matter. Appendix D provides short biographical sketches of the members of the responsible committee.

This report has been more than two years in the making. From the very beginning the committee attempted to extend its reach into the scientific community through both individual and general written invitations for contributions and through a number of public presentations of its progress.

Early on the COHS recognized the need to limit its scope to a manageable subset of the myriad issues and problems related to water. We have addressed ourselves solely to hydrologic science, omitting consideration of the applied forecasting (engineering hydrology) and management (water resources) aspects of water. **It is important to understand that the proposals here are not suggested as substitutes for or as pejorative reflections upon existing research or educational activities in the latter two fields but rather are intended as needed complementary, new initiatives.**

This report has undergone extensive review, both through the National Research Council's formal process and through various informal routes. We have tried to recognize all contributors in our acknowledgments (Appendix C).

The problem we address here is one within the infrastructure of science in the United States, and our support for this work has come solely from domestic organizations. For these reasons, and to keep the job manageable, we have not attempted the much larger task of assessing the status of the infrastructure of hydrologic science in other countries.

In summary, we believe this report presents sound arguments and broad areas of action for bringing identity and unity to hydrologic science. We must succeed in this endeavor or the field will complete its fragmentation, and the other geosciences will develop and subsume the parts they need. This would likely result in a failure to generate the necessary base of water science that is coherent and complete at the large space and time scales of emerging environmental problems.

PETER S. EAGLESON, *Chairman*
Committee on Opportunities in the
Hydrologic Sciences

Contents

Summary and Conclusions

SYNOPSIS

The drama of geophysics is now vividly transmitted by views of our living, changing planet from space platforms. This perspective has inspired multidisciplinary efforts to describe and understand the interactive functioning and continuing evolution of the earth's component parts. In turn, these efforts have brought a fuller appreciation of the central role that the global circulation of water plays in the interaction of the earth's solid surface with its atmosphere and ocean, particularly in regulating the physical climate systems and the biogeochemical cycles.

This realization of the importance of water to the earth system at geophysical space and time scales has profound implications for the research and educational infrastructure of hydrologic science. We cannot build the necessary scientific understanding of hydrology at a global scale from the traditional research and educational programs that have been designed to serve the pragmatic needs of the engineering community.

Investments in water resources management over the last century have helped provide the remarkable levels of public health and safety enjoyed by the urban populations of the developed world. While we have spent lavishly to cope with the scarcities and excesses of water and to ensure its potability, we have invested relatively little in the basic science underlying water's other roles in the planetary mechanisms. The committee believes that this science, hydrologic science, has a natural place as a geoscience alongside the atmospheric, ocean, and solid earth sciences; yet in the modern scientific establishment this niche is vacant. The supporting scientific infrastructure, including distinct educational programs, research grant programs, and research institutions, does not now exist for hydrologic science and must be put in place. This report presents the supporting arguments and recommends specific remedial actions.

WATER AND LIFE

Life arose in water and there began its evolution from the simple plants and animals that were virtually all water to humans, who by weight are approximately two-thirds water. Water has unique physical and chemical properties that enable it to play key roles in regulating the metabolism of plants, animals, and even our living planet.

• *Elixir of life*—The peculiar molecular structure of water makes it an almost universal solvent; no other liquid can dissolve such a wide variety of compounds. Because cell membranes are permeable only to certain dissolved substances, water is the elixir of life, essential— as blood and lymph—both for the nourishment of cells and for the removal of their wastes. It plays this same role at all higher levels of life's organization: for the individual plant or animal, the household, the city, civilization, and, apparently, for the earth itself.

• *Climatic thermostat*—A gram of water can absorb more heat for each degree of temperature rise than can most other substances. This high specific heat gives water a correspondingly large thermal inertia, making it the flywheel of the global heat engine. Because of water's special character, oceans and large lakes fluctuate little in temperature, and the heat-sensitive proteins within plant and animal cells are insulated by their aqueous baths.

• *Global heat exchanger*—When changing among its liquid, vapor, and solid states (at constant temperature), a gram of water absorbs or yields more heat than do most other substances. The phase changes of water on the earth are powered by the sun. Solar energy stored in water vapor as latent heat during evaporation travels with the vapor in the atmospheric circulation until it is released when the vapor condenses into precipitation. In this way both water and heat are redistributed globally.

The range of surface temperatures and pressures on the earth is such that water is plentiful in its life-supporting liquid state and yet moves freely and vigorously to its vapor and solid states as well. The more we learn about our desiccated, and apparently barren, neighboring planets, the more we wonder if our good fortune is not a result as well as the cause of life on the earth.

EARTH'S HYDROLOGIC CYCLE

The pathway of water as it moves in its various phases through the atmosphere, to the earth, over and through the land, to the ocean, and back to the atmosphere is known as the hydrologic cycle. This cycle is the framework of hydrologic science and occurs over a wide range of space and time scales.

In one round trip through this cycle a single water molecule may assume various roles: dissolving minerals from the soil and carrying them to nourish plants, quenching the thirst of humans, acting as a coolant, and serving as a solvent or chemical reactant in industrial processes. In any of these roles this water molecule may return to its hydrologic pathway in new chemical compounds or, along with its associates, it may be mixed with various solid and liquid substances. Thus the hydrologic cycle is not defined solely by the quantity of water moving through it but also by that water's quality.

Furthermore, many things affected by water in its relentlessly repetitive cycle have their own effects on that cycle. Prime examples are plants, which regulate the rate at which a land surface returns water vapor to the atmosphere, and humans, who alter nearly all aspects of water on land. Such interactions are not limited to living things, however, if we consider longer time scales. For example, alluvial aquifers, formed over geological time through erosion and sedimentation by glaciers and streams, form a dynamic component of the contemporary hydrologic system. Our water-based environment has arrived at its present state through eons of coevolution of climate, life, and the solid earth. The central role of water in the evolution and operation of the earth system provides a rationale for seeing hydrologic science as a geoscience of stature equal to that of the atmospheric, ocean, and solid earth sciences.

We now understand that the hand of mankind is altering the earth's environment on a global scale by virtue of such widespread activities as deforestation, urbanization, and pollution. These actions of humans now extend to the "ends of the earth": high latitudes, deserts, and mountains, where they affect sensitive environments and where hydrologic data and understanding are absent; they cause global-scale change in the hydrologic cycle. Ensuring the security of water supplies and protecting against flood, drought, and rising sea level require that we understand these changes. We must learn to incorporate human activity as an active component of the hydrologic cycle in all environments.

A DISTINCT GEOSCIENCE

Over the past 60 years, the evolution of hydrologic science has been in the direction of ever-increasing space and time scales, from small catchment to large river basin to the earth system, and from storm event to seasonal cycle to climatic trend. Hydrologic science should be viewed as a geoscience interactive on a wide range of space and time scales with the ocean, atmospheric, and solid earth sciences as well as with plant and animal sciences. To establish and retain the individuality of hydrologic science as a distinct geoscience, its domain is defined as follows:

• *Continental water processes*—the physical and chemical processes characterizing or driven by the cycling of continental water (solid, liquid, and vapor) at all scales (from the microprocesses of soil water to the global processes of hydroclimatology) as well as those biological processes that interact significantly with the water cycle.

(This restrictive treatment of biological processes is meant to include those that are an active part of the water cycle, such as vegetal transpiration and many human activities, but to exclude those that merely respond to water, such as the life cycle of aquatic organisms.)

• *Global water balance*—the spatial and temporal characteristics of the water balance (solid, liquid, and vapor) in all compartments of the global system: atmosphere, oceans, and continents.

(This includes water masses, residence times, interfacial fluxes, and pathways between the compartments. It does not include those physical, chemical, or biological processes internal to the atmosphere and ocean compartments.)

SOME UNSOLVED PROBLEMS

The enlarged scope of hydrologic science brings increased complexity and increased interaction with allied sciences. New questions of physical behavior arise, such as the following:

• How do we aggregate the dynamic behavior of hydrologic processes at various space and time scales in the presence of great natural heterogeneity?

• What are the feedback sensitivities of atmospheric dynamics and climate to changes in land surface hydrology, and how do these vary with season and geography?

- What can the soil, sediment, vegetation, and stream network geometry tell us about river basin history and about the expected hydrologic response to future climate change?
- What can we learn about the equilibrium and stability of moisture states and vegetation patterns? Is "chaotic" behavior a possibility?
- How are water, sediment, and nutrients exchanged between river channels and their floodplains?
- What are the states and the space-time variabilities of the global water reservoirs and their associated water fluxes?
- How can the necessary and fundamental links between the deterministic and stochastic models of rainfall fields be established?
- What are the physical factors that control the snow cover-climate feedback process and its role as an amplifier of climatic change?

Fundamental chemical and biological questions arise also that are often soil-related and hence at the other extreme of scale. A few examples will give the flavor:

- How can we employ modern geochemical techniques to trace water pathways, to understand the natural buffering of anthropogenic acids, and to reveal ancient hydroclimatology?
- What is the nature of the feedback processes that occur between biochemical processes and the various physical transport mechanisms in the soil?
- What is the relative importance of different flow paths and residence times to the chemistry of subsurface water?
- How much transfer of adsorbed materials from one grain to another occurs during streambed storage?
- How should we quantify the processes that determine the transport and fate of synthetic organic chemicals that enter the ground water system?

These and many other fundamental problems of hydrologic science must be addressed to provide the ingredients for solving the sharpening conflicts of humans and nature. Many, if not most, will require coordinated multidisciplinary field studies conducted at the appropriate scales. Others, such as the measurement of unknown oceanic precipitation and evaporation, will require sensors, often satellite-borne, that are still undeveloped. Progress in many areas of hydrologic science is currently limited by a lack of (high-quality) data.

DATA ISSUES

Hydrologic processes are highly variable in space and time, and this variability exists at all scales, from centimeters to continental scales, from minutes to years. Data collection over such a range of scales is difficult and expensive, and so hydrologic models usually conceptualize processes based on simple, often homogeneous, models of nature. This forced oversimplification is impeding both scientific understanding and management of resources.

In the history of the hydrologic sciences as in other sciences, most of the significant advances have resulted from new measurements, yet today there is a schism between data collectors and analysts. The pioneers of modern hydrology were active observers and measurers, yet now, designing and executing data collection programs (as distinct from field experiments with a specific research objective) are too often viewed as mundane or routine. It is therefore difficult for agencies and individuals to be doggedly persistent about the continuity of high-quality hydrologic data sets. In the excitement about glamorous scientific and social issues, the scientific community tends to allow data collection programs to erode.

Such programs provide the basis for understanding hydrologic systems and document changes in the regional and global environments. Modeling and data collection are not independent processes. Ideally, each drives and directs the other. Better models illuminate the type and quantity of data that are required to test hypotheses. Better data, in turn, permit the development of better and more complete models and new hypotheses. We must reemphasize the value and importance of observational and experimental skills.

EDUCATIONAL ISSUES

Higher education in hydrology, especially at the graduate level, has long been the province of engineering departments in most universities. Doctoral and master's degree programs administered by these departments have been directed toward the traditional concerns of water resources development, hazard mitigation, and water management as predicated on societal needs. The research focus in these departments has properly been the analysis and solution of problems related to engineering practice, on the premise that these problems contribute palpably to the technical knowledge base required for water resources allocation, the management of floods and droughts, and pollution control. Current societal needs, as expressed through legislative action or executive orders, are as important to the choice of research problems and their methods of solution as are the flow of scientific ideas and technological breakthroughs.

This well-developed and successful line of inquiry differs markedly from that pursued in the pure sciences, such as chemistry. The difference, in fact, is exactly analogous to that between the disciplines of chemistry and chemical engineering. Chemistry is the science that deals with the composition, structure, and properties of substances and the reactions that they undergo. Chemical engineering deals with the design, development, and application of manufacturing processes in which materials undergo changes in their properties. The first discipline is a science, dealing with puzzle solving (i.e., motivated by a question), whereas the second is an application of science, dealing with problem solving (i.e., motivated by the answer). Hydrology has a long and distinguished history of problem solving, but where is the antecedent science of puzzle solving?

The education of hydrologic scientists offers challenges as great as those in engineering hydrology, but the spirit of the enterprise is different, just as it is between education in chemistry and in chemical engineering. The choice of research problem is occasioned by its level of development within the hierarchy of the science, by the availability of new methods with which to solve it, and by the desire to understand a hydrologic phenomenon more deeply. The solution of the problem advances the development of the science and expands the conceptual framework that gives it meaning. It is this kind of internally driven intellectual pursuit that motivates the pure scientist and that must be instilled by the educational process that forms her or his professional outlook. That is the challenge to hydrologic science, and it differs from the challenge to engineering.

Graduate Education in the Hydrologic Sciences

As a result of this challenge, graduate education in the hydrologic sciences should be pursued independently of civil engineering. Some universities do this by housing "water science" programs in departments such as geography or geology. However, few offer a coherent program that treats hydrology as a separate geoscience. It is a premise of this report that hydrology—expanded in scope, importance, and potential—must escape mere inclusion as an option under engineering, geology, or natural resources programs. Establishment of specialized Ph.D. and master's degree programs is, therefore, necessary to enhance the identity of hydrology as an established science. Graduates are needed who are considered first and foremost as hydrologists, not as civil engineers or geologists who know something about hydrology.

Undergraduate Education in the Hydrologic Sciences

Few undergraduate programs exist in hydrology, and most professionals gain entry to the field from engineering or from the geosciences. However, the geosciences and civil engineering both have suffered a precipitous decline in undergraduate enrollment in recent years. Thus the hydrologic sciences face a potential recruitment problem created, at least in part, by the increasing difficulties students face as they enroll in courses in these majors, the primary obstacle being the required capabilities in physical science and mathematics.

The nearly universal demand for computer literacy has left students with little time for commitment to laboratory and field courses. The consequences of this are both profound and disturbing. Students (and faculty) have become separated from the physical world they seek to master. If a major rejuvenation of the "observational" components of higher education were to occur, it would serve to improve the quality of professionals entering hydrologic science and also perhaps to attract larger numbers of experientially motivated students to the field.

Science Education from Kindergarten Through High School

The discussion above makes clear that the success of graduate programs in the hydrologic sciences will depend on the quality of undergraduate preparation in pure science and mathematics, which, in turn, depends critically on the educational background obtained in precollegiate years. Like the statistics for geosciences and civil engineering majors, those for science education among high school students show a dismal trend. Less than 50 percent of high school graduates in the United States have completed more than one year of mathematics and one year of basic science. Less than 10 percent have taken a physics course.

SCIENTIFIC PRIORITIES

The diversity and range of scale of the frontier problems in hydrologic science are illustrated clearly by the examples listed earlier. The committee ranked a long list of these problems according to their expected contribution to scientific understanding under the premises that (1) the largest potential for such contribution lies at the least explored scales and in making the linkages across scales, and (2) hydrologic science is currently data-limited. From the most promising on this ranked list, allied problems were grouped into a small set of broader but unranked research areas of highest priority.

Priority Categories of Scientific Opportunity (Unranked)

The committee suggests the following five research areas as those now offering the greatest expected contribution to the understanding of hydrologic science.

- **Chemical and Biological Components of the Hydrologic Cycle**

In combination with components of the hydrologic cycle, aqueous geochemistry is the key to understanding many of the pathways of water through soil and rock, for revealing historical states having value in climate research, and for reconstructing the erosional history of continents. Together with the physics of flow in geologic media, aquatic chemistry and microbiology will reveal solute transformations, biogeochemical functioning, and the mechanisms for both contamination and purification of soils and water.

Water is the basis for much ecosystem structure, and many ecosystems are active participants in the hydrologic cycle. Understanding these interactions between ecosystems and the hydrologic cycle is essential to interpreting, forecasting, and even ameliorating global climate change.

- **Scaling of Dynamic Behavior**

In varied guises throughout hydrologic science we encounter questions concerning the quantitative relationship between the same process occurring at disparate spatial or temporal scales. Most frequently perhaps, these are problems of complex aggregation that are confounding our attempts to quantify predictions of large-scale hydrologic processes. The physics of a nonlinear process is well known under idealized, one-dimensional laboratory conditions, and we wish to quantify the process under the three-dimensional heterogeneity of natural systems, which are orders of magnitude larger in scale. This occurs in estimating the fluxes of moisture and heat across mesoscale land surfaces and in predicting the fluvial transport of a mixture of sediment grains in river valleys. It arises in attempting to extend tracer tests carried out over distances of 10 to 50 m in an aquifer to prediction of solute transport over distances of hundreds of meters to kilometers. It occurs in extrapolating measurements of medium properties in a small number of deep boreholes (as in the Continental Scientific Drilling Program) to characterize fluid fluxes at crustal depth.

The inverse problem, disaggregating conditions at large scale to obtain small-scale information, arises commonly in the parameterization of subgrid-scale processes in climate models and in inferring the subpixel properties of remote sensor images.

Solving these problems will require well-conceived field data collection programs in concert with analysis directed toward "renormalization" of the underlying dynamics. Success will bring to hydrologic science the power of generalization, with its dividends of insight and economy of effort.

• **Land Surface-Atmosphere Interactions**

Understanding the reciprocal influences between land surface processes and weather and climate is more than an interesting basic research question; it has become especially urgent because of accelerating human-induced changes in land surface characteristics in the United States and globally. The issues are important from the mesoscale upward to continental scales. Our knowledge of the time and space distributions of rainfall, soil moisture, ground water recharge, and evapotranspiration are remarkably inadequate, in part because historical data bases are point measurements from which we have attempted extrapolation to large-scale fields. Our knowledge of their variability, and of the sensitivity of local and regional climates to alterations in land surface properties, is especially poor.

The opportunity now exists for great progress on these issues for the following reasons. Remote sensing tools are available from aircraft and satellites for measurement of many land surface properties. Critical field experiments such as the completed First ISLSCP (International Satellite Land Surface Climatology Program) Field Experiment (FIFE) and the Hydrologic-Atmospheric Pilot Experiments-Modelisation du Bilan Hydrique (HAPEX-MOBILHY), and others under way and planned, promise to improve both measurement and understanding of hydrologic reservoirs and fluxes on several scales. Additional experiments in a range of environments are needed. Finally, numerical models exist that are capable of integrating results from regional and global measurement programs and focusing issues for future experiments.

• **Coordinated Global-scale Observation of Water
   Reservoirs and the Fluxes of Water and Energy**

Regional- and continental-scale water resources forecasts and many issues of global change depend for their resolution on a detailed understanding of the state and variability of the global water balance. Our current knowledge is spotty in its areal coverage; highly uneven in its quality; limited in character to the quantities of primary historical interest (namely precipitation, streamflow, and surface water reservoirs); and largely unavailable still as homogeneous, coordinated, global data sets. The World Climate Data Program (WCDP) has undertaken the considerable task of assembling the historical and current data, and the World Climate Research Program (WCRP) is planning the necessary global experimental program, the Global Energy and Water Cycle Experiment (GEWEX), to place future observations on a sound and coordinated scientific foundation. Many nations must contribute for this program to be successful. The United States should play a major role in GEWEX through the support of key experimental components and accompanying modeling efforts. Of particular importance in this regard is NASA's Earth Observing System (EOS) program, which will include observing and data systems as well as scientific experiments for multidisciplinary study of the earth as a system.

• Hydrologic Effects of Human Activity

For at least two decades hydrologists have acknowledged that humans are an active and increasingly significant component of the hydrologic cycle. Quantitative forecasts of anthropogenic hydrologic change are hampered, however, by their being largely indistinguishable from the temporal variability of the "natural" system.

Experiment and analysis need to be focused on this question. Identification of the signal of change within the background noise of spatial and temporal variability will require observations at regional scale and over many annual cycles. Forecasting the course of future change will be eased by understanding what changes have already occurred.

Data Requirements

• Maintenance of Continuous Long-Term Data Sets

The hydrologic sciences use data that are collected for operational purposes as well as those collected specifically for science. Improvements in the use of operational data require that special attention be given to the maintenance of continuous long-term data sets of established quality and reliability. Experience has shown that exciting scientific and social issues often lead to an erosion in the data collection programs that provide a basis for much of our understanding of hydrologic systems and that document changes in regional and global environments.

• Improved Information Management

The increasing emphasis on global-scale hydrology and the increasing importance of satellite and ground-based remote sensing lead to use of large volumes of data that are collected by many different agencies. An information management system is needed that would allow searching many data bases and integrating data collected at different scales and by different agencies.

• Interpretation of Remote Sensing Data

Effective use of remote sensing data is now too difficult for many hydrologic scientists, because the interpretation often depends on a detailed knowledge of sensor characteristics and electromagnetic properties of the surface and atmosphere. Hydrologic data products should be made available in a form such that scientists who are not remote sensing experts can easily use the information derived.

• Dissemination of Data from Multidisciplinary Experiments

Special integrated studies, such as HAPEX, FIFE, and GEWEX, that involve intensive data collection and investigation of the fluxes of water, energy, sediment, and various chemical species, produce high-quality data sets that have value lasting far beyond the duration of the experiment. Optimal use of these data requires broader and more timely distribution beyond the community of scientists who are involved in the experiments.

Education Requirements

• Multidisciplinary Graduate Education Program

The broad range of education inputs to graduate study in hydro-logic science necessitates the formation of a multidisciplinary pro-gram in the hydrologic sciences. This program should be either a department unit or a confederation of faculty from host departments that is assured of autonomy and resources by upper-level administration. The program would educate graduate students who are considered first and foremost as hydrologists, not geologists, geographers, or engineers who have some background in hydrology.

• Experience with Observation and Experimentation

The changing nature of hydrologic science requires the development of coordinated, multidisciplinary, large-scale field experiments. Graduate students should be given experience with modern observational equipment and technologies within their university programs, and mechanisms should be developed to facilitate their participation in these experiments, irrespective of their university of study. When the experiments are planned, the inclusion of a diverse array of studies should be an integral part of the plan. Undergraduate students of science should have experience with measurement of natural phenomena, preferably in field situations as well as in controlled laboratory settings.

• Visibility to Undergraduate Students

Programs should be developed to make hydrologic science more visible as a scientific discipline to undergraduate students. These programs should include such elements as research participation, internships at laboratories and institutes, curricula that introduce the latest innovations, visiting distinguished lecturers, media development, and in-service institutes for teachers.

RESOURCES AND STRATEGIES

Development of hydrology as a science is vital to the current effort to understand the interactive behavior of the earth system. Achieving this comprehensive understanding will require the kind of long-term disciplinary and interdisciplinary effort that can be sustained only by a vigorous scientific infrastructure. In conclusion this committee presents those resources and strategic actions that it believes are necessary to support a viable hydrologic science in the United States.

Resources

- **Research Grant Programs**

The central role of water in the earth system over a broad range of space and time scales provides the scientific rationale for a unified development of hydrologic science. The associated need to create and maintain a cadre of hydrologic scientists requires development of a focused image and identity for this science. Establishment of distinct but coordinated research grant programs in the hydrologic sciences would address both of these issues.

Support for research in hydrologic science in the United States is scattered among various agencies of the federal government. In keeping with the pragmatic origins of the science, the "action" agencies, such as the U.S. Geological Survey, the U.S. Environmental Protection Agency, the National Aeronautics and Space Administration, the National Weather Service, and the Agricultural Research Service, manage water-related research programs oriented to their own specific missions. The basic science fraction of this research, quite properly, is small in comparison with the applied. The amount of funds spent in-house is large with respect to external grants, and there is little coordination of effort at the interagency level.

Support for basic research in hydrologic science is concentrated within the National Science Foundation but is diffused there among the divisions of the Geosciences Directorate, each with a mandate oriented toward its own interests. This partitioning not only slights important hydrologic areas, such as aqueous chemistry and the earth's vegetation cover, but also ensures that there is no cultivation of a coherent research program in hydrologic science, and that the science achieves no established identity. A broad research grant program is needed that accommodates hydrology's natural role as a coupler of traditional disciplines, particularly solid earth science, atmospheric science, and terrestrial ecology.

• Fellowships, Internships, and Instructional Equipment

At the graduate level, this committee recommends establishment of special research fellowships in the hydrologic sciences. These should be designed to train students for research in a specific branch of hydrology and to increase the number of students equipped to investigate interdisciplinary problems. Travel fellowships will enable students to enroll in specific courses, to interact with key scientists, and to participate in large-scale, coordinated experiments. Fellowships are especially important in increasing participation by women, ethnic minorities, and the handicapped, as are internships for the retraining of mature scientists from allied disciplines.

At the undergraduate level, there is a strong need for providing modern, sensitive instructional equipment for students' use in the field and to back this up with logistical support for field trips and field classes.

Summer or academic year institutes for kindergarten through twelfth grade teachers can provide a basic science and mathematics background taught in the context of hydrology.

• Coordinated Field Experiments

Field studies involving multiple disciplines can often achieve more than the sum of their separate disciplinary goals by coordination of observations around a common, multidisciplinary objective.

Such coordinated field experiments include short-term, large-scale, multicollaborator studies, sometimes called campaigns or given acronyms such as GEWEX. These campaigns (e.g., the Global Atmosphere Research Program/Atlantic Tropical Experiment (GATE) and the Tropical Ocean and Global Atmosphere (TOGA) Program are widely used by the other geosciences, particularly to characterize mesoscale and larger phenomena, but are just coming into use in hydrology (e.g., HAPEX and FIFE). Such a program serves as an umbrella under which individual investigators carry out their work. Support of well-conceived campaigns is essential to the advancement of hydrologic science.

A second type of coordinated field experiment is the long-term or base-line study, such as the Long-Term Ecological Research (LTER) Program being carried out at specific sites under the leadership of the National Science Foundation. Formalized agency programs supporting faculty and student involvement in field experiments and instruction at these facilities are badly needed.

It should be the responsibility of universities and government agencies to inculcate the necessary planning and observational skills for all these modes of research through a steadfast, long-term commitment to the teaching and financial support of field work in the hydrologic sciences.

- **Long-Term Observations**

Continuous, long-term records of hydrologic-state variables (e.g., soil moisture, temperature, atmospheric humidity, and concentration of dissolved and suspended substances) and hydrologic fluxes (e.g., precipitation, streamflow, and evaporation) are essential, among other things, to quantify the variability of these quantities. These records have value in such areas as identification of global change (e.g., the Mauna Loa carbon dioxide record), isolation of mechanisms, and estimation of the risk of flood and drought.

The committee must renew the plea here for unwavering support of the collection and storage of long-term hydrologic records. These resources are like a patient's medical record: useless during apparent health, but invaluable when illness appears. The only certainty is that if records are not kept, they will not be available when needed.

- **Access to Data Bases**

The immediate, unrefined products of observation and experimentation are scientific data. These are obviously available to those who collect them, but their primary value is often realized by others at a later date and in a quite different scientific context. For hydrologic science to move forward it is essential that data sets, once acquired, be properly identified and described (i.e., purpose, location, instruments, spatial and temporal coverage, and so forth), be cataloged and archived (including archival maintenance), and be made available to the scientific community at reasonable cost and effort. Resources are needed for these tasks.

Strategies

To further the recognition and establishment of hydrologic science as a separate geoscience, hydrologists can take many actions, either individually or through their scientific societies. These include the following:

- *Make use of relevant scientific societies* as platforms for communication, advocacy, organization, and education.
- *Cultivate interest in hydrologic science* among the appropriate mission-oriented agencies of the federal government. There is a need

to argue for the allocation of a greater fraction of water research money to be spent by the agencies on basic hydrologic science. There is further need to seek interagency planning and coordination of how these monies are used.

• *Consider the establishment of a separate journal* for hydrologic science.

• *Stimulate joint meetings and symposia* among the relevant scientific societies concerning issues of hydrologic science in order to foster interdisciplinary understanding and cooperation.

• *Review, in five years, the progress toward achieving the goals* elaborated in this report, assessing the vitality of the field, surveying the changes that have occurred, and making recommendations for further action.

CONCLUSION

To meet emerging challenges to our environment we must devote more attention to the hydrologic science underlying water's geophysical and biogeochemical role in supporting life on the earth. The needed understanding will be built from long-term, large-scale coordinated data sets and, in a departure from current practices, it will be founded on a multidisciplinary education emphasizing the basic sciences. The supporting educational and research infrastructure must be put in place.

The benefits society will ultimately receive from a thorough scientific understanding of water behavior are many. Advances in the areas of irrigation, drinking water and ground water supplies, improved recreational areas and wildlife habitat, and flood and drought forecasting and planning are only a few examples. Improved hydrologic science will provide a foundation for decision making, resulting in protection and improved management of the world's water resources.

Water and Life

WONDROUS WATER

Life arose in water and there began its evolution from the simple plants and animals that were virtually all water to humans, who by weight are approximately two-thirds water. The average human contains nearly 50 liters of water and must replace about 5 percent of it daily for vital bodily functions.

Water has unique physical and chemical properties that enable it to play key roles in regulating the metabolism of our living planet.

• *Elixir of life*—The peculiar molecular structure of water makes it an almost universal solvent; no other liquid can dissolve such a wide variety of compounds. Because cell membranes are permeable only to certain dissolved substances, water is the elixir of life, essential— as blood and lymph—both for the nourishment of cells and for the removal of their wastes. It plays this same role at all higher levels of life's organization: for the individual plant or animal, the household, the city, civilization, and, apparently, for the earth itself.

• *Climatic thermostat*—A gram of water can absorb more heat for each degree of temperature rise than can most other substances. This high specific heat gives water a correspondingly large thermal inertia, making it the flywheel of the global heat engine. Because of water's special character, oceans and large lakes fluctuate little in temperature, and the heat-sensitive proteins within plant and animal cells are insulated by their aqueous baths.

• *Global heat exchanger*—When changing among its liquid, vapor, and solid states (at constant temperature), a gram of water absorbs or

17

yields more heat than do most other substances. The phase changes of water on the earth are powered by the sun. Solar energy stored in water vapor as latent heat during evaporation travels with the vapor in the atmospheric circulation until it is released when the vapor condenses into precipitation. In this way both water and heat are redistributed globally.

The range of surface temperatures and pressures on the earth is such that water is plentiful in its life-supporting liquid state and yet moves freely and vigorously to its vapor and solid states as well. The more we learn about our desiccated, and apparently barren, neighboring planets, the more we wonder if our good fortune is not a result as well as the cause of life on the earth.

ROUND AND ROUND AND ROUND IT GOES

The pathway of water as it moves in its various phases through the atmosphere, to the earth, over and through the land, to the ocean, and back to the atmosphere is known as the hydrologic cycle (Figure 1.1).

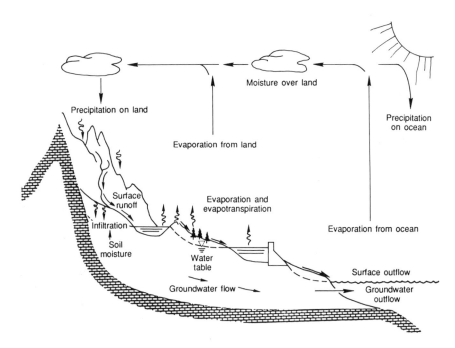

FIGURE 1.1 Elements of the hydrologic cycle. SOURCE: Reprinted, by permission, from Chow et al. (1988). Copyright © 1988 by McGraw-Hill, Inc.

In one round trip through this cycle a single water molecule may assume various roles: dissolving minerals from the soil and carrying them to nourish plants, quenching the thirst of humans, acting as a coolant, and serving as a solvent or chemical reactant in industrial processes. In any of these roles this water molecule may return to its hydrologic pathway in new chemical compounds or, along with its associates, it may be mixed with various solid and liquid substances. Thus the hydrologic cycle is not defined solely by the quantity of water moving through it but also by that water's quality.

Furthermore, many things affected by water in its relentlessly repetitive cycle have their own effects on that cycle. Prime examples are plants, which regulate the rate at which a land surface returns water vapor to the atmosphere, and humans, who alter nearly all aspects of water on land. Such interactions are not limited to living things, however, if we consider longer time scales. For example, alluvial aquifers, formed over geological time through erosion and sedimentation by glaciers and streams, form a dynamic component of the contemporary hydrologic system. Our water-based environment has arrived at its present state through eons of coevolution of climate, life, and the solid earth.

WATER AS ENABLER AND SUSTAINER OF CIVILIZATION

Water, so fundamental to maintaining life, was also critical to the development of civilization. In fact, it is not an exaggeration to state that civilization was born of water. Without access to and some degree of control over water, the coming of civilization would have been impossible. Water quenches people's thirst, supports their crops, and provides transportation and power. Water has long been essential to trade and to communication as well.

Water and Agriculture

The great early civilizations blossomed in the valleys of important river systems: the Nile of Egypt, the Tigris-Euphrates of Mesopotamia, the Indus of northern India (in what is now Pakistan), and the Hwang Ho of China. Generally, the Neolithic civilizations came into prominence some 10,000 years ago, when humans began mastering the skills and tools that gave them a measure of control over their environment. A critical step was the beginning of agriculture. As long as the efforts of virtually all the populace were required for subsistence food production, there was little time or energy for civilization to develop. However, as agriculture evolved and became more reliable and efficient, some people were freed from its burdens. They had time to make bricks and pottery, weave wool, cotton, and flax into fabrics, or work with metal. Trade developed. People congregated in villages and towns. In turn, other achievements ensued, such as the development of mathematics and writing.

Egypt provides a clear illustration of the relationship between a river and the development of an agricultural society. In late June, like clockwork, the lower Nile began to rise. By late September, the whole floodplain was covered. Then, as the waters receded, shrinking back to the main channel by October, they left behind a soil-building layer of silt, and a great agricultural potential. "Egypt," Herodotus said some 2,400 years ago, "is the gift of the river." The Egyptians used their hydraulic engineering skills to make the most of this already beneficial relationship: they built simple canals, dikes, and reservoirs to help manage the water and increase crop production.

Measurements of the Nile's water level were made at the second cataract as early as 1800 B.C. Two thousand years ago the Romans began to make regular measurements of the river's stages at Cairo, thus initiating the longest hydrologic record in the world. This "Nilometer" water-level gage was calibrated by the Roman naturalist Pliny the Elder (A.D. 23-79) in subjective terms that dramatically reveal the importance of the river to Egyptian life (Figure 1.2). As is often

FIGURE 1.2 Pliny the Elder's calibration of the Nile River's stages. SOURCE: Reprinted, by permission, from Dooge (1988). Copyright © 1988 by Blackwell Scientific Publishers Ltd.

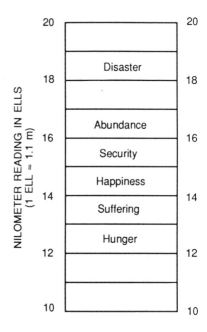

the case along uncontrolled rivers, social disaster accompanied both low stage (i.e., drought) and high stage (i.e., flood). Mesopotamia, part of a region often called the Fertile Crescent by archaeologists, consisted roughly of what is today Iraq and parts of Iran. Although millennia of climate change and human neglect have caused the region to become arid and inhospitable, this area that separates the valleys of the Tigris and the Euphrates rivers was once rich and productive. From 4000 B.C. the Sumerian civilization in the Tigris-Euphrates valley created an impressive irrigation system, including a great canal—the Nahrwan, about 120 m wide and over 320 km long—that fed smaller canals and channels throughout the valley. They also invented the water wheel to help transfer this water into ditches and furrows.

By 2400 B.C., however, the Sumerian culture was in decline, apparently for reasons directly related to the failure of its irrigated agriculture. The Sumerians had no drainage system to carry off excess water, and the salts left behind by evaporation of the irrigation water accumulated on the fields, rendering them unsuitable for growing crops.

Water and Climate Change

Climate is the fundamental determinant of water availability and hence of where humans have migrated and settled. Defined as the average local weather (i.e., temperature, pressure, precipitation, cloud cover, wind speed, and so on) over a long period of time (say, 30 years), climate is not a constant. It fluctuates with periodicities of 100,000, 41,000, and 21,000 years because of predictable variations in the earth's orbit, and it changes irregularly, for unknown reasons, on all time scales. These fluctuations influence both the amount and distribution of precipitation. A culture tied closely to a particular climate will be in jeopardy when its water supplies are reduced.

Water and the City

Whether village, town, or city, few human habitations have existed whose founding did not depend on the proximity of water. Often it was water for agriculture and drinking that pinpointed the exact spot for a settlement. The sites of villages are not typically fortuitous; people (either deliberately or spontaneously) choose sites with natural advantages, often water-related, such as defensibility or ease of transportation and communication. Towns often have been founded at some junction of physically contrasted zones. For instance, many coastal cities occur where goods are transferred from seagoing

THE BLACK PLAGUE

In the mid-fourteenth century in Europe, one of every four people died within a span of three years. The plague, or Black Death, was not, however, merely a health disaster. The root causes are now hypothesized to have been ecological crises, and water played a primary role. The plague had its devastating impact because environmental destruction had drastically reduced food production, which in turn had left the lower-class population chronically undernourished and particularly susceptible to disease.

The chain of events that foreshadowed the plague began with rapid population growth and deforestation. As the population in the Middle Ages grew, people moved into the forests, clearing more and more land for agriculture. The population of Europe doubled or tripled during this era, spurred by new agricultural techniques that increased the amount of food available. Pressures on natural resources mounted. Increasingly marginal lands were farmed, and shortly after 1300 the stress began to take its toll.

There is evidence that by 1300 large river systems were being severely harmed by poor land use practices. Food production on degraded farmlands dropped. Crops in fertile floodplains were increasingly lost to excessive flooding caused by rapid runoff from denuded slopes. Human wastes from the crowded cities polluted the waterways. Disease spread by contaminated water swept through urban areas. Climate played a major part in the scenario. Under the influence of the Little Ice Age, Europe became increasingly wet and cool, and this weather hastened the deterioration of grain crops, both before and after harvest. A population weakened by hunger and debilitated by disease was tragically susceptible to the plague. Thus, although the Black Plague was the outward manifestation of crisis, a major contributing cause was the degradation of Europe's water supplies.

Source: Francko and Wetzel (1983).

ships to land transport or river craft. For similar reasons cities are often found at the junctions of mountain and plain. Cities are born to take advantage of beneficial geography. Ancient seaports, for example, declined or were moved when effective access to the water was cut off by siltation, and modern seaports may disappear or be moved as a result of rising ocean levels.

WATER AS A HAZARD

Floods and droughts plagued humans even before they adopted an agricultural lifestyle.

Floods

A flood is an overflowing of water onto land not usually submerged. For scientific purposes the size of a flood is usually measured in terms of the maximum flow rate (cubic meters per second) of the flooding stream and depends on the rate, duration, and areal extent of the rainfall as well as on the nature and condition of the land on which it falls. For purposes of public safety the maximum water-surface elevation relative to the elevation of the bank is of prime concern, because what may be a flood at one place along a stream may be a well-controlled flow at another place.

Floods are a natural phenomenon important to the life cycle of many biota, not the least of which is mankind. Floods became a problem only as humans established farms and cities in the bottomlands of streams and rivers. In so doing they not only exposed their lives and property to the ravages of floods, but also exacerbated floods by paving the soil and constricting the stream channels. Over time continued urbanization of natural floodplains has caused great annual losses of both wealth (Figure 1.3) and human life (Figure 1.4). The Hwang Ho, for example, is sometimes known as China's Sorrow, a

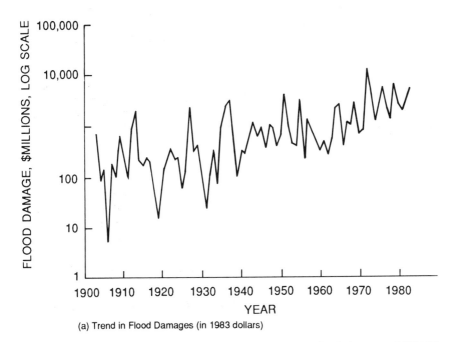

(a) Trend in Flood Damages (in 1983 dollars)

FIGURE 1.3 Historical trend in annual U.S. losses due to flood damage. SOURCE: Reprinted from Hudlow et al. (1984) courtesy of the National Weather Service.

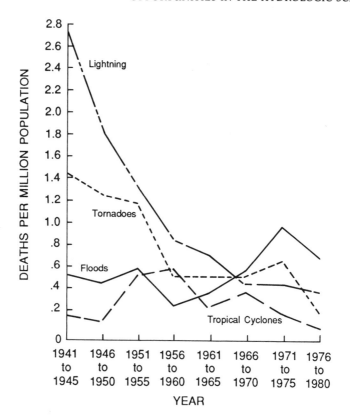

FIGURE 1.4 Population-adjusted death rates in the United States from four storm hazards, 1941 to 1980. SOURCE: STORM DATA, published monthly by the National Climatic Data Center. National Environmental Satellite, Data, and Information Series. National Oceanic and Atmospheric Administration, Asheville, N.C.

river so erratic and dangerous that a single flood reportedly has caused a million deaths. As early as the eighth century B.C., the Chinese were building dikes in attempts to confine the Hwang Ho's shifting channel and control its great destructive power. The Hwang Ho is also called the Mother of China because of the fresh topsoil the floods bring to the land.

The major scientific challenge with respect to floods here lies with improved short-range forecasting, but the principal hope for reduction of the losses lies with public policy that regulates development in the floodplain.

THE JOHNSTOWN FLOOD
MAY 31, 1889

When the storm struck western Pennsylvania it was the worst downpour that had ever been recorded for that section of the country. The Signal Service called it the most extensive rainfall of the century for so large an area and estimated that from six to eight inches of rain fell in twenty-four hours over nearly the entire central section. On the mountains there were places where the fall was ten inches. . . .

. . . When he arrived at the dam it was about ten minutes to three. There was no one actually out on the dam then, just at the ends, and the water was pouring over the breast. . . .

"It run over a short spell," he said, "and then about half of the roadway just fell down over the dam.

"And then it just cut through like a knife." . . .

When the dam let go, the lake seemed to leap into the valley like a living thing, "roaring like a mighty battle," one eyewitness would say. . . .

The water advanced like a tremendous wall. Giant chunks of the dam, fence posts, logs, boulders, whole trees, and the wreckage of the Fisher place were swept before it, driven along like an ugly grinder that kept building higher and higher. . . .

Most of the people in Johnstown never saw the water coming; they only heard it. . . .

. . . "a roar like thunder" was how they generally described it. . . .

Everyone heard shouting and screaming, the earsplitting crash of buildings going down, glass shattering, and the sides of houses ripping apart. . . .

. . . Once clear of the wireworks, the wave kept on coming straight toward him, heading for the very heart of the city. Stores, houses, trees, everything was going down in front of it, and the closer it came, the bigger it seemed to grow. . . .

The height of the wall was at least thirty-six feet at the center. . . .

. . . Now boxcars, factory roofs, trees, telegraph poles, hideous masses of barbed wire, hundreds of houses, many squashed beyond recognition, others still astonishingly intact, dead horses and cows, and hundreds of human beings, dead and alive, were driven against the bridge. . . .

SOURCE: Excerpts reprinted, by permission of Simon & Schuster, Inc., from McCullough (1968), pp. 21-149. Copyright © 1968 by David G. McCullough.

NOTE: This particular flood, which claimed a total of 2,209 people either dead or missing, was the result of an engineering design that failed to provide adequate emergency spillway capacity and/or management that failed to maintain the installed capacity of the submerged outlet. Excerpts from The Johnstown Flood are presented here to illustrate the destructive power of the extreme flash flood. Scientific advancement may help provide advance warning of the potential for such a disaster.

Droughts

Like beauty, drought is in the eye of the beholder (or the water user). It is commonly agreed that "drought" signifies an extended shortage of water, but there the agreement stops. The intended use defines what constitutes a "shortage," and the meaning of "extended" is subjective.

A water shortage arises when tropospheric circulation, which controls storm tracks, shifts in a way that makes rainstorms less likely at the location in question. A persistent lack of rainfall results presumably from inertia in the phenomena causing the shift in atmospheric circulation and may continue for several years or for a decade or more. Little is really known about the causes and persistence of drought, however, and this is a fruitful area for research.

Toxicity

The by-products of human activities must be disposed of as either liquid, solid, or gas within one of the compartments of the earth system. Many of these waste materials are harmful to human health, and traditionally, humans have disposed of them in streams, rivers, and lakes (from which they are ultimately transported to the oceans). Because these freshwater bodies have also served as prime sources for water supplies, they were the first to be protected by legislation mandating particular treatment of point sources of wastes before allowing their return to the hydrologic cycle.

We now realize that many of the waste materials thought to be "out of sight, out of mind" when disposed of within the earth are dissolved in ground water and hence reenter the hydrologic cycle as these waters rejoin the rivers. Regulatory action to control this practice is in its infancy, and cleanup of past damage is a difficult and costly task.

Other residuals are discharged to the atmosphere, from which many return to the land either in rainfall or snowfall or as dry deposition, and then go into aqueous solution. Regulation of this activity has yet to be effective.

WATER AS A RESOURCE TO BE MANAGED

History of Water Management in the United States

The management of water has proven to be critically important throughout the history of civilization, but the sophistication of these efforts and the pace of change have increased dramatically in the past few centuries. During the first 100 years of the European colonization of North America, each colonial household met its own needs for water and waste disposal. However, with increasing population growth in the early 1700s, the responsibility for some of these functions shifted to villages and then to cities, and the first water control structures were built to impound water for sawmills and gristmills and to divert water for municipal supplies. Still, water development was hydrologically, hydraulically, and structurally unsophisticated. The science and the engineering of water structures were based on short-term observations, trial and error, and rules of thumb. These early efforts may have been small scale and intended for local services only, but they provided water for consumption and power for the mills that were central to the economy of the colonies.

Beginning in the early 1800s, as the eastern United States was becoming more urbanized and as the West was beginning to be settled, the concept of larger, regionally oriented water projects took root. The nation built several major water reservoirs in the mid-1800s to supply eastern cities and to support western irrigation and hydraulic mining.

The stage had been set for a federal role in river-basin-scale water development when in 1824 the Congress provided its first appropriations to the U.S. Army Corps of Engineers for clearing snags and sandbars from the Mississippi and Ohio rivers. In the late 1800s, the U.S. Geological Survey was established to gather and develop the hydrologic information needed to complement the nation's water management efforts. By 1902, the U.S. Bureau of Reclamation had been established with a mission of developing irrigation water to help small farmers in their efforts to settle the West.

Subsequent decades found more and more roles for the federal government with respect to water resources. Beginning with the New Deal legislation of the 1930s, river basin water development was seen as a device for achieving social objectives such as the regional transfer of capital and people. The Army Corps of Engineers, the Bureau of Reclamation, the Soil Conservation Service, and other agencies engaged in a period of intense water resources management in response to various pieces of legislation intended to control floods, irrigate 17 western states, conserve soil and water, and provide navigable streams, hydroelectric power, water supplies, recreation, drainage, and other functions. The accomplishments during this extremely active period of development are impressive indeed. However, the environmental damage resulting from these projects is now also understood. For example, dams have interfered with the environment and life cycle of aquatic organisms and with the transport of river-borne supplies of sediments to replenish eroding beaches; irrigation has led to salination of agricultural land and to pollution of wetlands by leachates.

Many factors (such as increased environmental consciousness, shifted political priorities, and increased expectations of economic efficiency) contributed to ending this era of large-scale development, but there is no denying the significance of the period to the evolution of the science of hydrology and of related science and engineering disciplines. Efforts to manage water resources helped create a new, human-modified national landscape and, in the process, established humans as an inseparable part of the hydrologic cycle. The design, construction, and operation of these large water projects have furthered the development of many practical disciplines such as hydrologic engineering and water resources management. They have also stimulated increased interest in an understanding of the hydrologic science underlying these river-basin-scale projects.

Provision of Safe Drinking Water

The management of water resources—throughout the nation and the world—has had as a goal the availability of clean water for human consumption. It is the single greatest requirement for public health and a condition that is generally taken for granted in the United States and other industrialized nations. In the United States, water quality problems were recognized early in the twentieth century and were tackled by sanitary engineers, who devised treatment methods such as the simple addition of chlorine to kill infectious organisms.

But in developing countries, nearly 2 billion people (of a world population of approximately 5 billion) lack safe drinking water. Most of these people have no public water supply or wastewater disposal service; water-borne diseases are commonplace. Continued progress in providing safe drinking water will allow countries to focus on other water issues, such as irrigation for sustainable agriculture, and environmental problems, such as soil erosion, deforestation, and hazardous waste management.

Contemporary Water Resources Management Problems

Efforts to provide for the water-related needs of the world's populations have been impressive, but population growth and the changing demands of increasingly sophisticated societies have put unparalleled pressures on water resources. Today in the United States, and perhaps throughout the world, it seems that nearly every community faces some type of water crisis and that these crises are more technically, politically, and socially complex than those faced in the past. The

ABEL WOLMAN
(1892-1989)

In the course of an exceptionally long and active career, Abel Wolman may have done more than any single person to bring the benefits of hydrologic science to the people of the world.

Wolman was born in Baltimore, Maryland, on June 10, 1892, the fourth of six children of Polish immigrants. He received a Bachelor of Arts degree from the Johns Hopkins University in 1913 and hoped to become a physician. Instead, his family persuaded him to enroll in Johns Hopkins University's newly opened School of Engineering, where he received his Bachelor of Science in Engineering with the first graduating class in 1915. His professional career had already started; he had begun collecting water samples on the Potomac River for the U.S. Public Health Service in 1912. After a year he joined the Maryland Department of Health, beginning an association that lasted until 1939.

During his period of state employment he performed some of his most distinguished scientific research in water purification. When Wolman began this work, his own family practiced water purification by tying a piece of cheesecloth around the spigot in their home to filter out stones and dirt that flowed through the city's water supply. Not only was the quality of drinking water in general highly variable and questionable, but water supply sources and waste disposal sites were also frequently the same. Outbreaks of waterborne diseases struck Baltimore and other American cities with alarming regularity.

Wolman worked with chemist Linn H. Enslow in the Maryland Department of Health to perfect a method of purifying water with chlorine at filtration plants. Although the idea of using chlorine as a purifying agent was not new, procedures were crude and produced wildly fluctuating water products. Wolman and Enslow developed a chemical technique for determining how much chlorine should be mixed with any given source of water, taking into consideration bacterial content, acidity, and other factors related to taste and purity. That collaboration produced the gift of safe drinking water for millions of people around the world.

Wolman's capacity and enthusiasm carried him into national and international service for a period that spanned six decades. A member of the first delegation to the World Health Organization (WHO), he worked on water supply, wastewater, and water resources problems throughout the world with WHO and the Pan American Health Organization. A consultant to Sri Lanka (then Ceylon), Brazil, Ghana, India, and Taiwan, he also chaired the committee that planned the water system for the new state of Israel, and he helped Latin American nations develop ways to finance their water systems.

Abel Wolman was truly a man who transcended political and social boundaries and made the world a more livable place.

water system infrastructure that exists in most areas today is aging and is approaching the limits of its capacity to provide the services for which it was designed. The demand for water for many uses is increasing, and the supply may be decreasing in some regions. The introduction into the environment of exotic chemical products exacerbates this problem.

Treatment of municipal and industrial wastewater does not assure pollution-free streams. A much more confounding issue is water quality degradation from distributed (i.e., non-point) sources. Non-point pollution includes fertilizer and sediment runoff from agricultural fields, acid deposition from the atmosphere, detergents, oils, metals and fecal material carried in urban storm sewers, and other substances that cannot be identified as coming from a single source. This type of pollution can be particularly difficult to identify, let alone prevent, as is clearly the case at the Chesapeake Bay in the eastern United States. Boston, Massachusetts, provides another example of the growing problem of non-point pollution. For approximately a century, aqueducts have carried water from reservoirs in central Massachusetts through a system of holding lakes to the Boston metropolitan area for municipal use. Until now the quality of this water has been such that water treatment was unnecessary. Non-point pollution from development on lands tributary to the holding lakes is changing this situation.

Traditional challenges in irrigation management have included salinity, streamflow characteristics, storage requirements, and provision of means for conveyance and drainage. More recently, the discovery of toxic trace elements in the irrigation drainage water of the San Joaquin Valley of California and the mining of ground water (i.e., its exploitation beyond what is renewed) have had a serious impact on the future of irrigated agriculture in the United States. Ground water mining for irrigation and contamination by trace elements are also occurring in many other parts of the world.

These are but a few examples of contemporary water problems and the challenges they are now presenting to hydrologists and water managers.

Emerging Water-Related Problems

We now understand that the hand of mankind is altering the earth's environment on a global scale by virtue of such widespread activities as deforestation, urbanization, and pollution. Accompanying this environmental change is global-scale change in the hydrologic cycle. Ensuring the security of water supplies and protecting against flood, drought, and toxicity require that we understand these changes.

Humans are introducing into the air, soil, and water of our planet chemicals foreign to the evolutionary process that produced contemporary plant and animal life. To safeguard life we must understand the water pathways and aqueous processes to which these chemicals are subjected as they move through the earth system. To meet these and other emerging challenges we must devote more attention to the hydrologic science underlying water's geophysical and geochemical role in supporting life on the earth. The needed understanding will be built from long-term, large-scale coordinated data sets and, in a departure from current practices, it will be founded on a multidisciplinary education emphasizing the basic sciences. The benefits society will ultimately receive from a thorough scientific understanding of water behavior are many. Advances in the areas of irrigation, drinking water and ground water supplies, improved recreational areas and wildlife habitat, and flood and drought forecasting and planning are only a few examples. Improved hydrologic science will provide a foundation for decision making, resulting in protection and improved management of the world's water resources.

SOURCES AND SUGGESTED READING

Chow, V. T., D. R. Maidment, and L. W. Mays. 1988. Applied Hydrology. McGraw-Hill, New York.

Dooge, J. C. I. 1983. On the study of water. Hydrol. Sci. J. 28(1):23-48.

Dooge, J. C. I. 1988. Hydrology in perspective. Hydrol. Sci. J. 31(1):61-85.

Ford, E. C., W. L. Cowan, and H. N. Holtan. 1955. Floods—and a program to alleviate them. Pp. 171-176 in Water, the Yearbook of Agriculture 1955. U.S. Department of Agriculture.

Francko, D. A., and R. G. Wetzel. 1983. To Quench Our Thirst: The Present and Future Status of Freshwater Resources of the United States. The University of Michigan Press, Ann Arbor, 148 pp.

Hudlow et al. 1984. HYDRO Tech. Note 4. National Weather Service.

Langbein, W. B. 1981. A History of Research in the USGS/WRD. Pp. 18-27 in Water Resources Division Bulletin (Oct.-Dec.). U.S. Geological Survey.

McCullough, David G. 1968. The Johnstown Flood. Simon and Schuster, New York, 302 pp.

Schneider, S. H., and R. Londer. 1984. The Co-evolution of Climate and Life. Sierra Club Books, San Francisco, 563 pp.

2

The Hydrologic Sciences

Science draws its excitement from new and wondrous views of nature. By punching holes in the heavens, the telescope has revealed the immensity of the universe and thereby captured the imagination and support of the public for astronomy. At the other extreme, the microscope and the particle accelerator, through revelation of the structure of matter, have served the same purpose for materials science, molecular biology, and atomic physics.

The drama of geophysics is now vividly transmitted by views of our living, changing planet from space platforms. This perspective has inspired multidisciplinary efforts to describe and understand the interactive functioning and continuing evolution of the earth's component parts. In turn, these efforts have brought a fuller appreciation of the central role that the global circulation of water plays in the interaction of the earth's solid surface with its atmosphere and ocean, particularly in regulating the physical climate systems and the biogeochemical cycles.

The global distributions of rainfall, snowfall, evaporation, and accumulated surface and subsurface water affect the local extent and global distribution of biomass and biological productivity. Changes in land cover and biological productivity can, in turn, affect hydrologic processes on both local and global scales. Water exerts thermostatic control over local air temperature wherever evaporation or snow cover occur. Water movement couples the land with the oceans through the solution, entrainment, and transport of minerals and sediments;

32

both liquid water and ice are powerful agents of erosion and join with plate tectonics in shaping the land surface. This realization of the importance of water to the earth system at geophysical space and time scales has profound implications for the research and educational infrastructure of hydrologic science. We cannot build the necessary scientific understanding of hydrology at a global scale from the traditional research and educational programs that have been designed to serve the pragmatic needs of the engineering community.

THE UNIQUENESS OF WATER ON THE EARTH

The surface of the earth has abundant liquid water, yet our neighboring "terrestrial" planets—Venus and Mars—have little. Venus, Earth, and Mars all have atmospheres with clouds and solar-forced circulation. The primary constituents of the clouds—sulfuric acid on Venus, dust on Mars, and water on Earth—are markedly different, however. The major components of the earth's atmosphere (nitrogen and oxygen) are controlled by biological processes, whereas the Venusian and Martian atmospheres (both carbon dioxide) are governed by abiotic processes.

Theories for the uniqueness of water on the earth fall into two classes, genetic and evolutionary. The genetic theory holds that chemical equilibrium of accreting gas and dust in the solar nebula led to the formation of solid constituents richer in hydrated minerals at greater distance from the proto-sun. Once these minerals were incorporated into Venus, Earth, and Mars, their water and other volatiles were released to varying degrees over time in the formation of planetary atmospheres. The evolutionary theory, on the other hand, contends that the planetismals began with similar inventories of volatiles and that subsequent events, perhaps including meteoritic impacts, led to their current composition.

In any event, the higher accretion temperature and tectonic activity of Venus led to heavy outgassing, followed by irreversible photo-dissociation of any water into hydrogen, which escaped to space, and oxygen, which reacted with surface elements. The carbon dioxide remained to create a runaway greenhouse effect, resulting in Venus's current hot (464°C), dry surface.

On Mars the outgassing has been limited by lower accretion temperatures and by the absence of tectonic activity, but there is evidence of surface erosion by some flowing liquid, presumably water. The fate of this water is unknown. The (largely) carbon dioxide atmosphere of Mars is thin, and the current surface temperature is a cold

PALEOHYDROLOGY OF MARS

In 1972 the Mariner 9 spacecraft returned the first crude, remotely sensed images of channels and valleys on the planet Mars. Subsequently, in the late 1970s and early 1980s, the Viking mission spacecraft returned thousands of high-resolution pictures showing the apparently fluvial origin of these landforms. Mars had been thought to be a cold planet whose limited volatile components had remained frozen throughout its history. The fluvial channels and valleys showed that in its distant past Mars apparently had displayed a dynamic hydrologic cycle.

In the planetary sciences, comparisons and analogies function to generate the hypotheses that stimulate the building of new theory. We now know that at least one other planet besides the earth had an active system that generated runoff. On Mars the remnants of this hydrologic system display strange paradoxes relative to the familiar earth system. Martian channels are up to 100 km wide and 2,000 km long. They arise from source areas of collapsed terrain, indicating that fluid came to the surface from subsurface sources. Seeing the effects of large-scale exfiltration on Martian topography has given geomorphologists new ideas about the origin of landscapes on the earth in regions where emerging ground water rather than surface runoff is the dominant agent carving out river valleys.

Comparison of processes on the two planets should reveal much about the character of runoff production and its interaction with a planetary surface.

−53°C, leading to the presence of seasonal polar caps of frozen carbon dioxide and to speculation that there may be extensive subsurface frozen water.

It appears that the earth also once had a carbon dioxide atmosphere that was sharply reduced by some unique process (most probably biological), and that its water is the result of tectonically driven outgassing over geological time.

It is particularly important to the dynamics and energetics of the earth system that all three phases of water (solid, liquid, and vapor) coexist over the range of earthly temperatures and pressures. This is unique to the earth among the terrestrial planets (see the phase diagram of water given in Figure 2.1).

THE EARTH'S HYDROLOGIC CYCLE

Powered by the sun, the phase changes of water on the earth involve storage and release of latent heat that at once drive the atmospheric

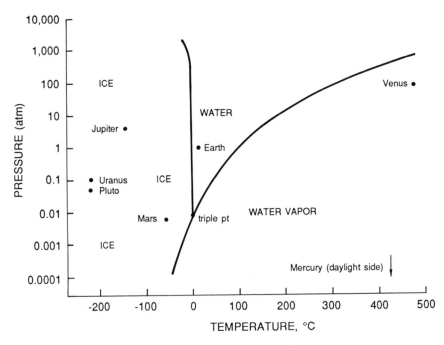

FIGURE 2.1 Planetary positions on the phase diagram of water.

circulation and redistribute both water and heat globally. Condensing in the atmosphere where it releases its latent heat, the liquid (or solid) water falls as precipitation, runs to the sea, and through evaporation regains its cargo of latent heat and returns through the atmosphere to wash the land again. This process is called the *hydrologic cycle*. It is the framework of hydrologic science (Figure 2.2) and occurs across a wide spectrum of space and time scales. It affects the global circulation of both atmosphere and ocean and hence is instrumental in shaping weather and climate. Water's efficiency as a solvent makes low-temperature geochemistry an intimate part of the hydrologic cycle. All water-soluble elements follow this cycle at least partially, or completely if they are in a chemical compound that is volatile as well as soluble. **The hydrologic cycle is thus the integrating process for the fluxes of water, energy, and the chemical elements.**

THE IMPORTANCE OF WATER ON THE EARTH

As far as is known, the earth alone supports life, and this life is active geophysically and geochemically as well as biologically. As

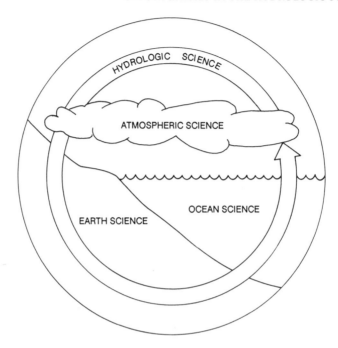

FIGURE 2.2 Hydrologic science is a geoscience.

examples, consider the role of biota in the cycling of oxygen, nitrogen, sulfur, and carbon on the earth:

• The oxygen in our atmosphere originally was released by plants after they began to evolve 2 billion to 3 billion years ago, and it is maintained by them through the photosynthetic decomposition of water molecules.

• Certain bacteria (as well as lightning and combustion) act to convert free atmospheric nitrogen into a chemical form that can be used by plants and animals.

• In the process of decomposing organic material in the sediments of swamps, marshes, and eutrophic lakes, bacteria use sulfates washed from the atmosphere by preciptation, reducing them to volatile sulfides that return to the atmosphere for reoxidation.

• Photosynthesis removes carbon dioxide from the atmosphere. Through respiration and decomposition plants and microorganisms pump carbon dioxide into the soil, where it joins with that washed out of the atmosphere by precipitation. Some of this carbon dioxide reacts with rock minerals and forms various carbonates and bicarbonates. These are dissolved in ground water and carried to the oceans, where they join other carbon compounds entering directly from the atmosphere. By this mechanism primitive biota reduced the carbon dioxide concentration of the earth's early atmosphere. Contemporary life forms still remove carbon dioxide in this way, and by incorporating it into plant material. However, humans are changing this natural balance by accelerating the release of carbon dioxide through combustion, and by modifying photosynthesis through deforestation.

In the oceans carbonaceous minerals are incorporated into animal shells and then are deposited in ocean sediments; ultimately they become part of the earth's crust. These sedimentary rocks may eventually become involved in volcanism, whereby their carbon dioxide (as well as water entrapped in the marine sediments) is once more released to the atmosphere.

The earth is unique among the terrestrial planets both because it has an active water cycle and because it supports life. The water cycle is an essential part of the planet's life support mechanism and, to the extent that the biota are responsible for the earth's moderate surface temperatures, the biota permit the water cycle to exist. This synergism couples the animate and inanimate components of the earth into an evolving system. The central role of water in the evolution and operation of the earth system provides a rationale for seeing hydrologic science as a geoscience whose stature equals that of the ocean, atmospheric, and solid earth sciences.

EARLY SCIENTIFIC INSIGHTS

Concern for water as both a necessity of life and a possible hazard has been with humans throughout their existence. Drought and flood have driven the search for an understanding of water since the first civilizations formed along the banks of rivers. Pioneering hydraulic engineers built primitive dams, channels, and levees to meet their practical needs. They sought to understand the vagaries of water only when these engineering measures were insufficient.

Not until the rise of Hellenic civilization, about 600 B.C., did man attempt to understand nature just for the sake of understanding. Early Greek philosophers such as Herodotus, Hippocrates, Plato, and Aristotle theorized about the source of rivers and rain but failed to develop a complete understanding of the hydrologic cycle.

Precipitation was first measured in the fourth century B.C. by Kautilya of India, and streamflow was monitored by Hero of Alexandria in the first century A.D., but little further advance in understanding occurred until the Renaissance. Bernard Palissy (1510-1590), a French potter and naturalist who used his own field observations to build revolutionary theories, was the first to state categorically (and counter to Plato and Aristotle) that rivers can have no other source than rainfall. He gave the first correct explanation of the hydrologic cycle for temperate regions.

The first comprehensive hydrologic field study, in which Palissy's explanation was proved, was conducted by Pierre Perrault (1608-1680), a French lawyer and member of a distinguished family who became a hydrologist only after his dismissal (for embezzlement) as Receiver-General of Paris. Edmund Halley (1656-1742), an English Astronomer Royal, conducted experiments on evaporation after having vapor condensation interfere with his celestial observations on clear nights. He calculated that ocean evaporation is sufficient to replenish the inflowing rivers. With this proof the hydrologic cycle was firmly established, marking the beginning of scientific hydrology. During this early phase of investigation, hydrology received attention as a natural science worthy of study in its own right without concern for utility.

THE AGE OF APPLICATIONS

For the rest of the seventeenth and eighteenth centuries, Europe was transformed by the Industrial Revolution and the urbanization that accompanied it. Physicists and members of distinguished families lost interest in hydrology, and its development was left to engineers concerned with the urgent matters of water supply and sanitation. The subfield of hydraulics received great impetus then from civil engineers such as Henri Pitot, Antoine de Chézy, and G. B. Venturi, who were concerned with water supply and water power. The primary refinement of the hydrologic cycle during this period was provided by Jean-Claude de la Méthérie (1743-1817), who explained that rainfall has three possible fates: (1) direct movement to streams, (2) evaporation or transpiration and moistening of the soil, and (3) deep percolation to feed springs.

Water-related science developed spottily in response to the needs of engineering practice. Water scientists and engineers focused their attention on drainage basins commonly having a characteristic horizontal scale of 10 to 100 km. Because the early foundations of hydrologic science were built on experience with the middle latitudes, some inadvertent and long-lived biases were established. At middle latitudes, the atmospheric processes driving catchment hydrology are dominated by cyclonic motions having horizontal scales (e.g., 1,000 km) orders of magnitude larger than those of the individual catchments being studied. This disparity of scales encouraged the convenient assumption that the catchment is a passive participant in the hydrologic cycle, producing no feedback to the atmosphere from either its surface state or its streamflow. In addition, the hydrologic processes peculiar to the nondeveloping desert and to tropical and cold regions received little or no attention. In this context, T. J. Mulvaney (1822-1892), an Irish engineer, was apparently the first to deal with the unsteady rainfall-runoff problem in a catchment. His landmark 1851 work relating precipitation and the resulting maximum flood discharge opened a field of study that preoccupied applied hydrologists for the next 125 years. Applying higher mathematics for the first time in hydrology, Philipp Forchheimer (1852-1933) in Germany and C. S. Slichter (1864-1946) in the United States founded an elegant theory that describes ground water flow.

Until late in the nineteenth century hydrologic research in the United States remained the province of enterprising professors, inventors, prospectors, and wealthy amateurs. However, at that time the growing data needs of water management projects on large rivers (e.g., 100 to 1,000 km) led to the establishment of new public agencies, both federal and state, including the U.S. Weather Bureau and the U.S. Geological Survey (USGS). Government was now the prime mover in water research in the United States, although primarily in support of the practical missions of its agencies. The USGS, for example, was founded in 1879 to produce maps and data of a geological nature about the "products of the national domain."

The early history of the USGS embodies the development of hydrologic science in the United States, particularly in such areas as sediment transport (led by G. K. Gilbert), ground water (C. S. Slichter, O. E. Meinzer, and C. V. Theis), and water chemistry. In surface water, private consulting engineers such as R. E. Horton, Allen Hazen, L. K. Sherman, and Adolf Meyer remained at the forefront of research well into the twentieth century. O. E. Meinzer, as head of the ground water group of the USGS, brought together in the 1920s a group of geologists and hydraulic engineers to develop quantitative methods for the study of ground water. Out of this group came pioneering work on unsteady ground water flows by such leaders as C. V. Theis and C. E. Jacob.

This period of federal and state agency dominance of hydrologic science in the United States ended with the completion, in the 1950s, of the water projects delayed by World War II, and with the concurrent rise of both the environmental movement and the culture of government-supported university research.

In the early twentieth century the first English language textbooks on hydrology were published by Daniel Mead (in 1904) and Adolf Meyer (in 1919). These authors were engineers, and the hydrology subjects introduced into U.S. universities using these books were taught largely in departments of civil engineering, where the focus was naturally on questions of surface runoff, water supply, and floods.

THE STRUGGLE FOR SCIENTIFIC RECOGNITION

Although the International Union of Geodesy and Geophysics (IUGG) formed an International Commission on Glaciers in 1897, the first formal recognition of the scientific status of hydrology was the formation of the Section of Scientific Hydrology within the IUGG at its Rome assembly in 1922. Two years later, at the 1924 Madrid assembly, this new section established a commission on statistics charged with bringing uniformity into the publication of hydrologic information by the national services—a sign of the interests of the era.

In 1922 the U.S. National Research Council's Committee for the IUGG was called the American Geophysical Union (AGU). Its delegate to the Rome assembly returned with a recommendation that a new section of scientific hydrology be added to the AGU. The fate of this proposal illustrates the status of hydrologic science within the U.S. scientific community during the first half of the twentieth century. Despite repeated recommendations by ad hoc committees and biennial pleas from the IUGG, the leadership of AGU maintained for eight years that active scientific interest in the United States did not justify a separate section of scientific hydrology within the AGU.

Finally, in reviewing plans for transforming the AGU from a committee of the National Research Council into an independent society, approval for the new section was given. On November 15, 1930, the Section of Hydrology of the AGU came into existence with O. E. Meinzer as chairman and R. E. Horton as vice-chairman. At the next annual meeting of the AGU (April-May 1931), Horton presented a comprehensive analysis of the field, scope, and status of the science of hydrology as seen at that time (Horton, 1931). His definition of hydrology as a science was as follows:

> As a pure science, hydrology deals with the natural occurrence, distribution, and circulation of water on, in, and over the surface of the earth. . . . More specifically, the field of hydrology, treated as a pure science, is to trace out and account for the phenomena of the hydrologic cycle. (p. 190)

In defining the scope of hydrologic science, Horton went on to say:

> Both the scope and problems of hydrology are closely related to the various branches of applied hydrology. This is natural since it is mainly in the applications that new problems arise and the scope of the science is extended. . . . Its scope is limited to considerably less than the entire field of water science. (p. 191)

ROBERT E. HORTON
(1875-1945)

Robert E. Horton is often called the father of American hydrology because of the breadth and interdisciplinary nature of his contributions. He was a practical man who spent his professional life working as a hydraulic engineer for various government agencies and then as a private consultant in the northeastern United States. From his problem-solving experience, Horton persistently distilled general principles about processes in the hydrologic cycle. By promoting quantitative and mathematical approaches to the study of processes, and by emphasizing and illustrating the use of carefully measured data, he founded by example the science of analytical hydrology. Horton emphasized the need for "research to provide connective tissue between related problems."

Horton was a prolific author, known not only for the impressive volume of his work (his contributions did not cease with his death—posthumous papers continued to be published until 1949) but also for its high quality and authority. He could address diverse fields—hydraulics, meteorology, soil physics, and geomorphology—with equal skill. He even published a volume of short stories. He is remembered first, however, as a scientist of great vision, originality, and curiosity.

Horton is best known for his theory relating the infiltration capacity of soils to the generation of floods by surface runoff. He went on to emphasize the influence of soils and vegetation on runoff processes. His theory also provided a basis for analyzing the mechanics of soil erosion and for devising rational strategies for soil conservation. Throughout his career, Horton returned to the issue of flood generation, analyzing the physics and temporal variability of rainfall, the role of vegetation in controlling interception as well as transpiration and antecedent soil moisture, and the role of stream channels.

In a 1945 landmark paper, Horton provided a synthesis of the relations between runoff processes (as influenced by precipitation, soil, and vegetation) and the mechanics of hillslope and channel erosion, and thus of drainage basin development. That paper, summarizing the significance of his career-long scientific interest, continues to provide an inspiration for much of hydrology and geomorphology.

In using the hydrologic cycle to define the processes encompassed by hydrologic science, Horton recognized the diversity of scales by stating:

Any natural exposed surface may be considered as a unit area on which the hydrologic cycle operates. This includes, for example, an isolated tree, even a single leaf or twig of a growing plant, the roof of a building, the drainage basin of a river-system or any of its tributaries, an undrained glacial depression, a swamp, a glacier, a polar ice-cap, a group of sand dunes, a desert playa, a lake, an ocean, or the Earth as a whole. (p. 192)

OSCAR E. MEINZER
(1876-1948)

Oscar E. Meinzer is best known for his leadership at the U.S. Geological Survey (USGS). His career with the USGS spanned 40 years (1906-1946), when hydrogeology first gained recognition as a significant branch of earth science. Under Meinzer's leadership, a systematic scientific approach was applied to the problems of hydrogeology, and the underlying principles were defined. His work in codifying and organizing elements of the many disciplines related to the study of ground water helped define the fundamental concepts and underlying principles of a new science called ground water hydrology. His early involvement in studying the principles and occurrence of ground water in the United States earned him the moniker "father of modern ground water hydrology."

Meinzer was born on a farm near Davis, Illinois, in 1876. He graduated magna cum laude from Beloit College in 1901 and received his Ph.D. in geology from the University of Chicago in 1922. He joined the USGS as an aide in 1906 and in 1912 succeeded W. C. Mendenhall as chief of the Division of Ground Water.

His long term as ground water chief, from 1912 to his retirement in 1946, often is referred to as "the Meinzer years." It was a period of rich intellectual achievement during which researchers developed the fundamental blocks of knowledge that would support the new science.

In 1923 Meinzer published two reports that formalized the status of ground water hydrology as a science and that provided a state-of-the-art review of the ground water field. In aggregate, these benchmark reports, "The Occurrence of Ground Water in the United States, with a Discussion of Principles" and "Outline of Ground Water Hydrology, with Definitions" defined the breadth, scope, and philosophy of the new science. They provided a generation of ground water hydrologists with clear guidance and high standards for the conduct of ground water investigations.

Throughout Horton's definitive work it is apparent that

1. the central focus was still on the conservation of water mass at the scale of the river basin, where evaporation was characterized as a "water loss";

2. concern was exclusively with the physics of the hydrologic cycle, omitting any mention of chemistry and biology (with the exception of trees and vegetation); and

3. the qualifier "natural" was used to exclude concern with the effects of humans.

These restrictions reflect the continued dominance of the field in the United States by the engineering concerns of nation building. In the meantime, beginning in 1926, the National Research Council had undertaken the preparation of a series of volumes on the physics of the earth intended "to give the reader, presumably a scientist but not a specialist in the subject, an idea of its present status, together with a forward-looking summary of its outstanding problems." In 1936, upon the recommendation of the AGU, the National Research Council appointed a subcommittee on hydrology with O. E. Meinzer (geologist in charge, Division of Ground Water, USGS) as chairman to prepare a volume on hydrology as a conclusion to this series.

Meinzer opened that volume with a definition that somewhat tentatively proclaimed hydrology to be an earth science, but one with traditional methods and problems that set it apart (Meinzer, 1942)! His definition went beyond Horton's, however, to incorporate water-related chemical and biological activity:

Hydrology is, etymologically, the science that relates to water. It is, however, an earth science. It is concerned with the occurrence of water in this earth, its physical and chemical reactions with the rest of the earth and its relation to the life of the earth. It includes the description of the earth with respect to its waters. It is not concerned primarily with the physical and chemical properties of the substance known as water. Like geology and the other earth sciences, it uses the basic sciences as its tools, but in doing so, it has developed a technique and subject matter that are distinct from those of the basic sciences. (p. 1)

THE MODERN AGE OF HYDROLOGIC SCIENCE

By the mid-1900s, research on the scientific aspects of hydrology was well under way in university and government laboratories focused on understanding the laboratory-scale physical processes of the hydrologic cycle. Within the United States concern was beginning to arise about the quality of water and about the preservation of natural environments. These expanded interests found expression in a rephrasing of Meinzer's definition of hydrology as a science. The Ad Hoc Panel on Hydrology (1962) of the U.S. Federal Council for Science and Technology, chaired by Walter B. Langbein, offered the following explanation:

Hydrology is the science that treats of the waters of the Earth, their occurrence, circulation, and distribution, their chemical and physical properties, and their reaction with their environment, including their relation to living things. The domain of hydrology embraces the full life history of water on the Earth. (p. 2)

WALTER B. LANGBEIN
(1907-1982)

In his long and distinguished career, Walter B. Langbein never failed to direct his scientific endeavors to the public good. He was indeed a scientific public servant.

Langbein was born in Newark, New Jersey, on October 17, 1907. While attending evening classes at Cooper Union, he worked with a construction company in New York City. He received his Bachelor of Science degree in civil engineering in 1931 and he remained with the company until 1935, when he took a position as hydraulic engineer with the U.S. Geological Survey in Albany, New York.

Langbein's interest ran the full gamut of hydrology. He was recognized in the scientific community as among the world's most versatile, competent, and distinguished hydrologists. He was an innovator and leader in a diverse list of hydrologic subjects: floods and flood hydrology, unit hydrograph studies, soil temperature, lake and river sedimentation, closed lakes, evaporation, reservoir storage, river meanders, the hydraulic geometry of river channels, base level in physiography, and the theory of kinematic waves applied to riverbed form. His studies of flood frequency, translated to indicate flood risk, became the basis of the 1968 National Flood Insurance Act. He contributed significantly to our understanding of the role of land use in conjunction with climate in the evolution of river morphology in time and space. Through his participation in the programs of the International Hydrologic Decade (1965-1974), which he was instrumental in founding, Langbein focused attention on the problem of determining the worth of hydrologic data relative to water resources development, thereby pointing the way to rational design of hydrologic data collection systems.

For his scientific endeavors, pursued so cooperatively, Langbein was awarded the Bowie and Horton Medals of the American Geophysical Union, the J. C. Stevens Award of the American Society of Civil Engineers, the Distinguished Service Award of the U.S. Department of the Interior, and the Warren Prize of the National Academy of Sciences. He and Professor Korzun of the Soviet Union were co-recipients of the International Prize in Hydrology, awarded by the International Association of Hydrologic Sciences.

Langbein took his work seriously, yet possessed a dry sense of humor. He once remarked that one's professional career is a race against obsolescence. Any hydrologic judge would claim that Walter B. Langbein clearly won the race.

The panel observed further that hydrology is an interdisciplinary science, involving an integration of other earth sciences to the extent that these help to explain the life history of water and its chemical, physical, and biological constituents. This definition maintains the breadth and scientific flavor of the Meinzer version, but it retains almost exclusive explicit concern with what happens to the water.

Recognizing the need for international cooperation to effectively use transnational water resources and for broad-scale international cooperation to acquire hydrologic data, the United Nations sponsored the International Hydrologic Decade from 1965 to 1974. Perhaps the primary benefit of this program was a raising of consciousness about regional- and global-scale problems and about human impact on the hydrologic cycle. Consideration of the temperate latitude megalopolis and tropical latitude deforestation made it clear that the lateral scales of human alteration of the land surface were becoming commensurate with those of atmospheric moisture exchange. The feedback effects of land surface state on the influential atmospheric processes had acquired a practical importance, and it was realized that the necessary science base was missing.

It was the dramatic color photographs of the earth in space, however, that crystallized active interest in the interconnectedness of nature and in the changes being wrought by humans. This realization has found its way into contemporary views of the interactive role of man in the hydrologic cycle (Figure 2.3). It accepts that human activity has become an integral and inseparable part of the hydrologic cycle and that the quality of the water as it moves through the cycle is no less a concern than the quantity. In fact, the quality of water can even influence important quantitative fluxes of the hydrologic cycle.

STATUS OF UNDERSTANDING

Reservoirs and Fluxes of Water

The hydrologic cycle is illustrated as a global geophysical process in Figure 2.4. Little is known about the amount of water in the two limiting reservoirs, space and the earth's mantle, but there is evidence that they both exchange water with the primary crustal, ice, atmospheric, and oceanic reservoirs. The hydrogen in water is lost to space very slowly through diffusion of water vapor and methane molecules into the upper atmosphere and the subsequent escape of hydrogen atoms freed from these molecules by photochemistry. This effective water loss (indicated by the symbol L in Figure 2.4) occurs at a rate probably on the order of 10^{-4} km^3 per year. Addition of water

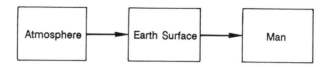

a. Classical Viewpoint

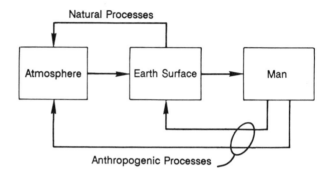

b. Modern Viewpoint

FIGURE 2.3 The role of man in the hydrologic cycle. SOURCE: National Research Council (1982).

(A) from space is controversial—perhaps icy comets are a source. Without doubt, however, volcanic activity continuously vents water vapor to the atmosphere (V_ℓ) and liquid water to the oceans (V_o). Again, the rate is uncertain, perhaps less than 1 km^3 annually. Water also recirculates on a geological time scale by the subduction (S) of water-containing crustal material in a tectonic hydrologic cycle, as illustrated in Figure 2.5.

More than 97 percent of all water in the land-ocean-atmosphere system is saline and resides in the oceans. This enormous reservoir is involved in processes of exchange with water on and under the land and in the atmosphere on many different space and time scales. The atmosphere, although it supports a global average precipitation rate (P) on the order of 1 m per year, contains only 0.001 percent of the earth's water at any moment. This is enough to cover the globe to a depth of about 2.7 cm. The atmospheric water storage is replenished once every 9 to 10 days (on the average) through evaporation (E). Evaporation, vapor transport in the atmosphere, condensation,

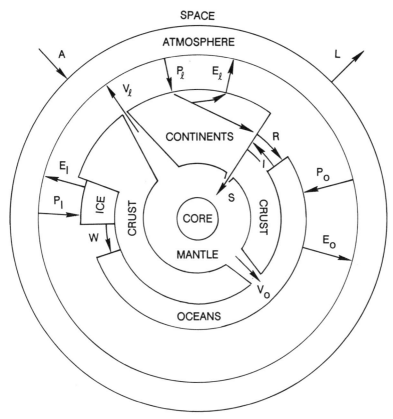

A = Additions of water from space
E_o = Evaporation from oceans
E_I = Evaporation (i.e., sublimation) from ice
E_ℓ = Evapotranspiration from land
I = Intrusion of seawater into continental aquifers
L = Loss of water to space
P_o = Precipitation on oceans

P_I = Precipitation on ice
P_ℓ = Precipitation on land
R = Runoff from continents
S = Subduction of water-containing crust
V_o = Volcanic venting to oceans
V_ℓ = Volcanic venting to atmosphere
W = Wastage of ice sheets to ocean

FIGURE 2.4 The hydrologic cycle as a global geophysical process. Enclosed areas represent storage reservoirs for the earth's water, and the arrows designate the transfer fluxes between them.

and precipitation are the fundamental mechanisms for distillation and redistribution of the earth's fresh water in a climatic hydrologic cycle. This cycle, usually referred to simply as the hydrologic cycle, plays a major role in the redistribution of incoming solar energy as well. It is illustrated quantitatively at global scale in Figure 2.6. About one-half of all solar energy reaching the earth's surface is used

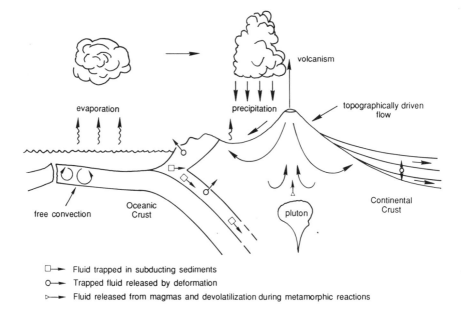

FIGURE 2.5 The tectonic hydrologic cycle. SOURCE: Reprinted from Forster and Smith (1990) courtesy of the Mineralogical Association of Canada.

in evaporation, of which 90 percent comes from the ocean (E_o) and 10 percent from the land (E_ℓ). The latent heat required for the phase change is carried with the resulting vapor by the wind until it is released when and where the vapor condenses. Water vapor is the most important of the greenhouse gases, acting to regulate the earth's surface temperature by absorbing and returning to the earth much of the thermal radiation emitted there.

Oceanic precipitation (P_o) and evaporation (E_o) have not been observed systematically because of obvious experimental difficulties. Their global average difference usually is estimated as the closure for a global water balance. Regional differences are important because they enhance or suppress thermohaline circulation. The distribution in space and time of storm precipitation is poorly understood in relation to the mechanics of storm genesis even for storms over land, where observation is much simpler.

The strength and the horizontal scale of this evaporation-precipitation cycle and associated energy redistribution are known to be highly variable with both season and geography but have not been well studied. In tropical regions, where thermal convection is a dominant atmospheric mechanism, as much as 50 percent of local precipitation may be derived from local evaporation. In the temperate latitudes with cyclonic atmospheric motions, it may be only 10 percent. **It is important to learn how patterns of surface wetness, temperature, reflectivity, and vegetation influence the formation of clouds and precipitation on a wide range of space and time scales.**

As a global average, only about 57 percent of the precipitation that falls on the land (P_ℓ) returns directly to the atmosphere (E_ℓ) without reaching the ocean. The remainder is runoff (R), which finds its way to the sea primarily by rivers but also through subsurface (ground water) movement, and by the calving of icebergs from glaciers and ice shelves (W). In this gravitationally powered runoff process, the water may spend time in one or more natural storage reservoirs such as snow, glaciers, ice sheets, lakes, streams, soils and sediments, vegetation, and rock. Evaporation from these reservoirs short-circuits the global hydrologic cycle into subcycles with a broad spectrum of scale. The runoff is perhaps the best known element of the global hydrologic cycle, but even this is subject to significant uncertainty. For example, the direct ground water flow to the sea is missing from

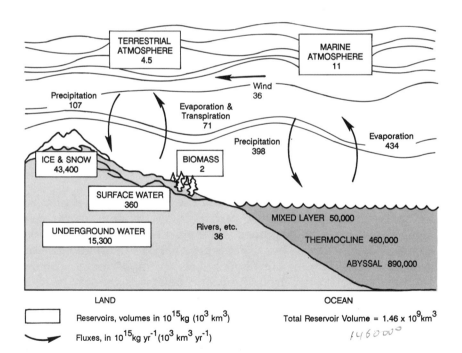

FIGURE 2.6 The hydrologic cycle at global scale. SOURCE: National Research Council (1986).

streamflow observations and may be as much as 10 percent of total continental runoff.

The water in the uppermost 1 or 2 m of the earth's crust is known as soil moisture and is a key determinant of the state of the earth system. With typical residence times of days to months, it provides the driving potential for moisture fluxes in the soil: upward to the roots of plants and as evaporation, and downward in recharging ground water. It serves the same storage and driving functions for dissolved chemical species. Both the carrier and the solutes interact with the soil medium.

It is important to learn the rates and pathways of moisture through the soil in order to predict soil chemical reactions, solute responses, and water quality changes. Soil moisture also plays an important role in soil formation—directly, through chemical weathering of rock, and indirectly, through life support of soil biota. Soil moisture also is an active agent in the depletion of soil minerals through leaching, and it influences the resistance of soils to erosion. Finally, it has an obvious and direct effect on the growth of vegetation, an aspect of the hydrologic cycle that has been intensively studied at the microscale because of its importance to agriculture. However, the heterogeneity of the subsurface medium presents many unsolved problems in understanding large-scale soil moisture behavior. For example, is it possible to infer the spatial structure of soil hydrologic properties from the large-scale geological processes responsible for rock and hence soil composition?

The porous earth and rock materials deep beneath the land surface and oceans constitute a great water reservoir. It has been estimated that some permeability must exist within the crust to depths of 13 to 20 km. Water filling these pores is called ground water, the permeable strata are called aquifers, and the upper limit of the saturated zone approximates what is called the water table.

Most movement of ground water is the result of topographic relief, and rates of flow are slow, perhaps a few centimeters per day. Depending on how far the ground water must travel to reach a surface discharge area, water in the shallow zone may remain underground for periods ranging from a few hours to more than 100 years. Among the important unsolved ground water problems is the fate of toxic elements and compounds in these waters. **It is necessary to understand the advection and dispersion of solutes and their reaction with the porous medium as well as the transport of microparticles and their filtration by the medium.** Water at great depth may take tens or even hundreds of thousands of years to pass through the subsurface, and often such water is highly mineralized. Evidence exists for circulation of ground water at depths of from 10 to 15 km, probably as a result of tectonically induced pressure gradients. However, it is not yet known how to measure and characterize the transport properties of these fractured rock masses.

The volume of ground water in the upper kilometer of the continental crust is an order of magnitude larger than the combined volume of water in all rivers and lakes and is equivalent to the total of all ground water recharge for about the last 150 years. The total volume of ground water is equal to almost one-fourth of all the nonoceanic water on the earth.

Discharges of ground water at topographically lower elevations of the earth's surface enable streams to flow even during prolonged rainless periods and after winter snows have melted. Coastal ground water reservoirs may be recharged locally by intrusion (I) of salt water from the ocean.

A major problem that recurs throughout geophysics is the representation of spatially aggregated nonlinear behavior in the presence of large spatial variability. In other words, given the dynamics at microscale, how can behavior at the macroscale be represented? This scale-transfer problem arises in the hydrologic sciences in attempts to describe the coupled fluxes of heat and moisture across large land surface elements, to couple the microscopic molecular processes of chemical reactions to the macroscopic averages of ground water transport equations, and to establish appropriate parameters for use in describing the behavior of ground water plumes at field scale.

The largest masses of fresh water exist as ice in Antarctica's and Greenland's ice sheets. This ice, with an average residence time of about 10,000 years, participates very slowly in the hydrologic cycle. However, the reservoirs are so large that small-percentage changes in ice volume can cause major changes in sea level on time scales of 100 to 10,000 years. Wastage (W) occurs primarily by melting or calving around the periphery. The Greenland ice sheet, if it melted, would yield enough water to maintain the flow of the Mississippi River for more than 4,700 years, and this ice sheet represents only 10 percent of the total volume of icecaps and glaciers. The greatest single item in the water budget of the earth, aside from the oceans, is the Antarctic ice sheet. It contains about 64 percent of all nonoceanic waters. Melting of just the small West Antarctic ice sheet would raise global sea level by about 7 m! The stability of those portions of the Antarctic ice sheet that are grounded below sea level, such as the West Antarctic ice sheet, is a major unsolved problem. A sudden slide of this sheet would cause sudden and calamitous sea level rise.

The addition or subtraction of ice from smaller icecaps and moun-

TABLE 2.1 Typical Residence Times of Global Water

Compartment	Typical Residence Time
Deep ground water	10^4 years
Icecaps and glaciers	10^4 years
Oceans	10^3 years
Shallow ground water	10^2 years
Soil moisture	10^2 days
Seasonal snow	10^2 days
Rivers	10 days
Atmosphere	10 days
Plants	10 days

NOTE: These residence times are averages computed, for atmosphere and ocean, by dividing the total water mass in the entire reservoir by the current rate of either inflow or outflow. For the ground water, "shallow" refers to the topmost kilometer of the continental crust; the associated water mass and flux are poorly known and thus so is the residence time. The residence time for deep ground water is a crude estimate for depths of 1 to 10 km.

SOURCES: Dooge (1984); Philip (1978); UNESCO (1971).

tain glaciers, reacting on time scales of 10 to 100 years, may appreciably affect the flow of certain rivers. The melting of ice in these glaciers appears to have caused one-third to one-half of the sea level rise observed in the past century. Snow cover on land and sea ice on the ocean vary rapidly and seasonally and exert a major influence on the earth's radiation budget and on the circulation of both the atmosphere and the ocean. **Particularly needed are new observational techniques to study and monitor the rates of snow accumulation and snow and ice melt over remote areas.** Because of the sensitivity of snow and ice reservoirs to climate change, it is important to monitor closely the extent of snow cover, the mass balances of mountain glaciers and ice sheets, and the West Antarctic ice sheet with its fringing ice shelves.

The magnitudes of these major water storages and fluxes are shown in Figure 2.6, and typical residence times are summarized in Table 2.1. How humbling it is to realize that despite man's seeming importance, his existence is made possible by the less than 1 percent of the earth's water that is directly available as fresh water.

Flux of Sediments

Water moves around on the earth's surface, and the large-scale primary driver is gravity. In its relentless downhill course to the sea, water sculpts the landscape through the processes of erosion, transport, and deposition. In so doing, it plays the key role in a geological-scale tectonic-climatic feedback system. Tectonic and volcanic processes lift the crust, creating the gradients that drive erosion, which in turn gradually reduces the gradients (Figure 2.7). If the elevation changes are large enough they may affect climate, and the erosion may be self-limiting by precipitation reduction as well.

In the process of shaping the landscape, runoff forms a tree-like network of channels into which the flow becomes concentrated. Although empirical "laws" describing the two-dimensional geometry of these networks have existed for almost half a century, **there is little quantitative understanding of the dynamics of channel formation or of the causal relationship between the three-dimensional network structure and the precipitation driving the erosion.** Such understanding would reveal fundamental scaling relationships of surface water hydrology over a broad range of spatial scales (i.e., 1 to 10^6 km^2) and would have immediate applicability to flash flood forecasting in ungaged watersheds and to parameterization of hydrologic processes in regional and global models. It would also help answer many fundamental geological questions about landscape formation.

In spite of decades of study, continuing deficiencies in our understanding of fluid turbulence seriously impair our ability to specify the relative rate of transport of various sizes of grains and aggregates

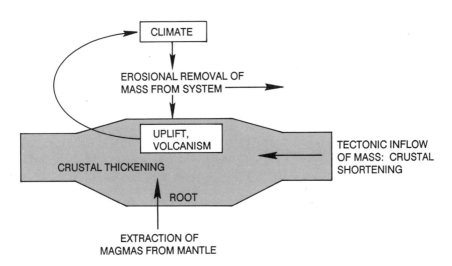

FIGURE 2.7 The tectonic-climatic feedback loop. SOURCE: Courtesy of B. L. Isacks and the Cornell Andes Project.

TABLE 2.2 Continental Yield of Water, Water-Borne Sediments, and Dissolved Solids

Continent	Area (10^6 km^2)	Annual Water Yield (10^3 km^3)	Annual Sediment Yield (10^{12} kg)	Dissolved Solids Concentration (ppm)	Annual Dissolved Load (10^{12} kg)
Africa	30.26	4.1	0.48	121	0.50
Asia	43.25	13.2	14.53	142	1.87
Australia	7.70	2.3	0.21	59	0.14
Europe	10.36	3.0	0.30	182	0.55
N. America	23.31	6.7	1.78	142	0.95
S. America	17.82	11.2	1.09	69	0.77

SOURCES: Dooge et al. (1973); Dooge (1984).

in streams and the duration of their storage at various locations within the channel system. This is important not only for its contribution to understanding erosion and deposition but also because many pollutants are moved through the system by their being adsorbed on sediment particles.

Table 2.2 contains the best current estimates of sediment fluxes from the earth's continents.

Flux of Dissolved Solids

Water is the universal solvent, and as it moves through the hydrologic cycle it dissolves and transports in solution solids as well as gases. Rain falling onto soil surfaces contains various gases and solids in solution. As water infiltrates the soil and moves downward, it picks up carbon dioxide from the soil, exchanges solutes with soil and rock particles, and becomes less acidic. The percolating waters convey their solute load through the ground water and into streams. This water has a dissolved-mineral signature that is dependent on the subsurface materials' properties, the flow path, and biological processes that recycle minerals. Knowledge of these solutes and their chemical kinetics can be used in tracer studies of subsurface water flow paths and to understand the rates of continental degradation and soil formation. Estimates of the total dissolved-solids runoff from the continents are given in Table 2.2.

Involvement of Biota

Water supports a variety of living organisms, and some have major interactions with the hydrologic cycle. The thin soil cover, for instance, is a result of physicochemical weathering by water and supports a vegetation cover that constitutes about 99 percent of the earth's terrestrial biomass. Through transpiration, the vegetation is responsible for most of the water returned directly to the atmosphere from the land, and the associated latent heat transfer is a major regulator of land surface temperature. Through photosynthesis the vegetation extracts carbon dioxide (a so-called greenhouse gas) from the atmosphere. The removal of vegetation modifies runoff and, in certain climates, may reduce local precipitation due to earth-atmosphere coupling.

The physical relationships among climate, soil, and vegetation that determine the dominance and stability of specific vegetation types at particular geographic locations are largely unknown, but understanding such relationships is necessary to anticipating the effects of climate change.

The soil also supports a variety of microorganisms that act on complex organic materials and reduce them to simpler organic compounds and ultimately to mineral form. Bacteria and fungi are particularly important in the carbon cycle and in regulating the availability of phosphorus, nitrogen, and sulfur. Their action results in the production of carbon dioxide and the development of humus, the organic detritus of decayed vegetation. Humus affects the infiltration and water-holding capacity of the soil and its resistance to erosion. Microorganisms also are responsible for transforming atmospheric nitrogen to a form usable by and essential to plants. While these well-known organisms are active near the soil surface, recent investigations have identified thousands of bacterial species up to 250 m below the surface. It is interesting to speculate on their natural role, if any, in the hydrologic cycle and on their possible use in the degradation of anthropogenic contaminants.

Oceanic microorganisms are responsible for roughly one-half of the earth's photosynthetic activity and therefore play a major role in atmospheric chemistry and the chemical quality of precipitation.

Wetlands are a primary source of atmospheric methane (another greenhouse gas) and perform a host of other hydrologic and biogeochemical functions. Serious scientific study of this complex biome is in its infancy, however.

In the last 500 years the hand of the human animal has been increasingly felt on the hydrologic cycle. Energy production, farming, urbanization, and technology have altered the albedo of the earth, the composition of its soil and water, the chemistry of its air, the amount of its forest, and the structure and diversity of the global ecosystem. These actions of humans now extend to the "ends of the earth"—high latitudes, deserts, and mountains, where they affect sensitive environments and where hydrologic data and understanding are absent. We must learn to incorporate human activity as an active component of the hydrologic cycle in all environments.

Summary

The evolution of hydrologic science has been in the direction of ever-increasing scale, from small catchment to large river basin to the earth system, and from storm event to seasonal cycle to climatic trend. Inevitably, increased scale brings increased complexity and increased interaction with allied sciences. New questions arise, such as the following:

• How do we aggregate the dynamic behavior of hydrologic processes at various space and time scales in the presence of great natural heterogeneity? This fundamental statistical-dynamical problem of hydrology remains unsolved.
• How can we employ modern geochemical techniques to trace water pathways, to understand the natural buffering of anthropogenic acids, and to reveal ancient hydroclimatology?
• What can the soil, sediment, vegetation, and stream network geometry tell us about river basin history and about the expected hydrologic response to future climate change?
• What can we learn about the equilibrium and stability of moisture states and vegetation patterns? Is "chaotic" behavior a possibility?

These and many other fundamental problems of hydrology must be addressed to provide the ingredients for solving the sharpening conflicts of humans and nature. Many, if not most, will require coordinated multidisciplinary field studies conducted at the appropriate scales. Others, such as the measurement of unknown oceanic precipitation and evaporation, will require sensors, often satellite-borne, that are still undeveloped. **Progress in many areas of hydrologic science is currently limited by a lack of (high-quality) data.**

HYDROLOGIC SCIENCE AS A
DISTINCT GEOSCIENCE

In 1931 Horton identified a hydrologic science of limited scope that was motivated by engineering practice to understand the quantity and movement of water at the small catchment scale. Over the next 30 years, first Meinzer (1942) and later Langbein's committee (Ad Hoc Panel on Hydrology, 1962) expanded the scope to embrace the full life history of water on the earth, including its chemical properties and its relation to living things. That definition needs modification now to reflect an evolving understanding of the science and to specify clear administrative boundaries for the science.

How have our perceptions of hydrology changed? **Hydrologic science can now be seen as a geoscience interactive on a wide range of space and time scales with the ocean, atmospheric, and solid earth sciences as well as with plant and animal sciences.** The new perceptions concern the interaction of the components and the range of scales.

Our perceptions of the necessary administrative boundaries also have changed. The ubiquity of water on the earth and its indispensability to life do not make hydrologic science out of all geoscience and biology. **Forging a separate identity for hydrologic science requires specifying and claiming its central elements, and locating its administrative boundaries as a flexible compromise between precedent and scientific completeness.** The scope of hydrologic science does not involve developing the physics, chemistry, and biology of water *within* the ocean and atmosphere reservoirs, for these processes are firmly in the recognized domains of the sibling geosciences. Such clear precedents do not apply to characterizing the lake and ice reservoirs, however. Limnologists and glaciologists are divided over their administrative homes. Indeed, in the United States the central scientific society for limnologists (particularly those with biological interests) is the American Society of Limnology and Oceanography (ASLO), and within the National Science Foundation (NSF) their primary research home is in the Division of Ocean Sciences. Many physical and some chemical limnologists consider themselves hydrologists, however, arguing that lakes are merely wide places in rivers. Glaciologists deal with a wide variety of snow and ice problems. Melting glaciers and snow cover, frozen ground, and lake and river ice have been a traditional part of hydrology, while glacial dynamics, large ice sheets, sea ice, and snow avalanches have not. In developing this report, the committee included the study of snow, ice, and lakes within its definition of hydrologic science.

To establish and retain the individuality of hydrologic science as a distinct geoscience, its domain is defined as follows:

• *Continental water processes*—the physical and chemical processes characterizing or driven by the cycling of continental water (solid, liquid, and vapor) at all scales (from the microprocesses of soil water to the global processes of hydroclimatology) as well as those biological processes that interact significantly with the water cycle.

(This restrictive treatment of biological processes is meant to include those that are an active part of the water cycle, such as vegetal transpiration and many human activities, but to exclude those that merely respond to water, such as the life cycle of aquatic organisms.)

• *Global water balance*—the spatial and temporal characteristics of the water balance (solid, liquid, and vapor) in all compartments of the global system: atmosphere, oceans, and continents.

(This includes water masses, residence times, interfacial fluxes, and pathways between the compartments. It does not include those physical, chemical, or biological processes internal to the atmosphere and ocean compartments.)

These boundaries are illustrated in Figure 2.8, and the range of process scales is shown in Figure 2.9.

The complex problems of global change illuminate the multidisciplinary nature of hydrologic science and make clear the need for extensive cross-discipline interaction in education as well as in research. The problems we face pay no attention to organizational boundaries; thus there are major areas of overlapping interest with other sciences and frequent needs to blur the stated boundaries. This is both inevitable and desirable. One example is the problem of the variability in space and time of storm precipitation wherein the search for hydrologic generalization demands incorporation of considerable atmospheric dynamics and thermodynamics. Similar trespassing must occur in

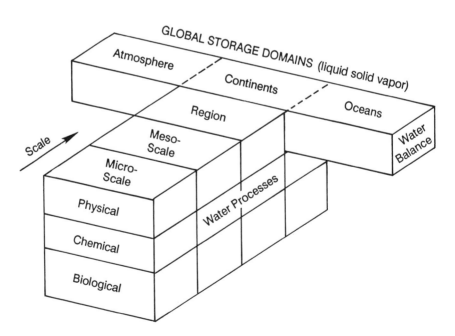

FIGURE 2.8 Hydrologic science: a distinct geoscience.

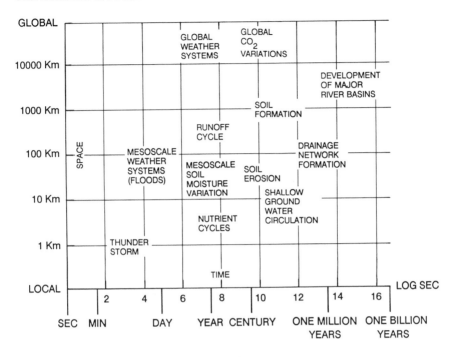

FIGURE 2.9 Illustrative range of process scales.

the areas of fluvial geomorphology, micrometeorology, and plant ecology (to name but a few) because of the importance to the hydrologic cycle of related processes such as erosion, energy flux, and transpiration.

The recent past has seen events that highlight the need for a separate and strong science of hydrology:

• pressure for economic development in the more extreme climates of the world such as the tropics, deserts, and arctic and alpine regions;
• realization of the capacity of humans to alter the hydrologic cycle on all scales, including a global scale; and
• evidence that anthropogenic changes to the chemistry of the earth's water are having harmful effects on the health of many humans.

Thus the science of hydrology has come to encompass a mix of natural and altered physical, chemical, and biological systems as well as to include important interactions with the engineering and social sciences. There is little doubt that coping with these issues in a timely fashion will require a much-improved scientific understanding of the earth system and its component parts. Unified and coherent treatment of hydrologic science is central to this larger effort.

SOURCES AND SUGGESTED READING

Ackermann, W. C. 1969. Hydrology becomes water science. Trans. AGU 50 (April):76-79.

Ad Hoc Panel on Hydrology. 1962. Scientific Hydrology. U.S. Federal Council for Science and Technology, Washington, D.C., 37 pp.

Biswas, Asit K. 1972. History of Hydrology. North Holland, Amsterdam, 336 pp.

Deevey, E. S., Jr. 1970. Mineral cycles. Sci. Am. 223(3):148-158.

Dooge, J. C. I. 1984. The waters of the Earth. Hydrol. Sci. J. 29(2):149-176.

Dooge, J. C. I. 1988. Hydrology in perspective. Hydrol. Sci. J. 33(1):61-85.

Dooge, J. C. I., A. B. Costin, and H. J. Finkel. 1973. Man's Influence on the Hydrological Cycle. Irrigation and Drainage Paper. Special Issue 17. Food and Agriculture Organization of the United Nations, Rome, 71 pp.

Eagleson, P. S. 1982. Hydrology and climate. Pp. 31-40 in Scientific Basis of Water-Resource Management. National Research Council, National Academy Press, Washington, D.C.

Forster, C., and L. Smith. 1990. Fluid flow in tectonic regimes. Fluids in Tectonically Active Regimes of the Continental Crust. Mineralogical Association of Canada, in press.

Horton, R. E. 1931. The field, scope, and status of the science of hydrology. Pp. 189-202 in Trans. AGU, Reports and Papers, Hydrology. National Research Council, Washington, D.C.

Jones, P. B., G. D. Walker, R. W. Harden, and L. L. McDaniels. 1963. The Development of the Science of Hydrology. Circular No. 63-03. Texas Water Commission, 35 pp.

Kasting, J. F., O. B. Toon, and J. B. Pollack. 1988. How climate evolved on the terrestrial planets. Sci. Am. 261(February):90-97.

Kerr, R. A. 1988a. In search of elusive little comets. Science 240:1403-1404.

Kerr, R. A. 1988b. Comets were a clerical error. Science 241:532.

Kerr, R. A. 1989. Double exposures reveal mini-comets? Science 243:13.

Langbein, W. B. 1981. A history of research in the USGS/WRD. Pp. 18-27 in Water Resources Division Bulletin (Oct.-Dec.). U.S. Geological Survey.

Livingstone, D. A. 1964. Chemical composition of rivers and lakes. U.S. Geological Survey Professional Paper 440 G.

Lovelock, J. E. 1979. Gaia. Oxford University Press, New York, 157 pp.

Meier, M. F. 1983. Snow and ice in a changing hydrological world. Hydrol. Sci. J. 28(1):3-22.

Meinzer, O. E., ed. 1942. Hydrology. Physics of the Earth—IX. McGraw-Hill, New York. (Republished by Dover, Mineola, N.Y.)

National Aeronautics and Space Administration Advisory Council, Earth System Sciences Committee. 1986. Earth System Science—Overview. NASA, Washington, D.C., 48 pp.

National Research Council. 1982. Scientific Basis of Water-Resource Management. National Academy Press, Washington, D.C., 127 pp.

National Research Council. 1986. Global Change in the Geosphere-Biosphere. National Academy Press, Washington, D.C., 91 pp.

Philip, J. R. 1978. Water on earth. Pp. 35-59 in Water: Planets, Plants and People. A. K. McIntyre, ed. Australian Academy of Science, Canberra.

Price, W. E., Jr., and L. A. Heindl. 1968. What is hydrology? Trans. AGU 49(2):529-533.

Prinn, R. G., and B. Fegley, Jr. 1987. The atmospheres of Venus, Earth, and Mars: a critical comparison. Annu. Rev. Earth Planet. Sci. 15:171-172.

Rainwater, F. H., and W. F. White. 1958. The solusphere—its inferences and study. Geochim. Cosmochim. Acta 14:244-249.

Schneider, S., and R. Londer. The Co-Evolution of Climate and Life. Sierra Club Books, San Francisco, 563 pp.

United Nations Educational, Scientific, and Cultural Organization (UNESCO). 1971. Scientific Framework of the World Water Balance. Technical Papers in Hydrology 7. UNESCO, Paris.

U.S. Geological Survey. 1968. Water of the World. USGS 0-288-962. U.S. Government Printing Office, Washington, D.C.

3

Some Critical and Emerging Areas

OVERVIEW

Earth processes are driven by two engines. The sun maintains the external engine that is responsible for the weather, surface erosion, and most oceanic processes. Radioactivity and primordial heat drive the internal engine that maintains the dynamic plate system and creates global topography. Hydrologic science plays a fundamental role in key mechanisms by which the external and internal engines make the earth such a singular planet.

This chapter presents some critical and emerging areas in hydrologic science. It is not exhaustive; the intention is to convey the flavor of the challenges and frontiers that make hydrology so critical a field of study in understanding the earth system. Toward this goal, the connection between hydrology and the earth's internal engine is explored. It is precisely through hydrologic processes that some of the most important interactions between the internal and the external engines occur.

The tectonic system and the hydrologic system come together in the earth's rigid outer skin, mainly in the upper 10 km of the continental crust. Hydrologic processes play an important role in the tectonic system; for example, subsurface waters, in responding to changing thermal and stress conditions, can have a significant impact on the mechanics of earthquakes. The evolution of sedimentary basins and the genesis of ore deposits are fundamentally influenced by ground water flows operating on time scales of 10^2 to 10^6 years and spatial scales of tens to hundreds of kilometers. The vastness of the scales involved brings enormous variability in the properties of the physical system, and new models to understand transport processes and their media of occurrence are being explored.

The subsurface is also where one of the major environmental impacts of human activities takes place. This is the deposition of different types of waste and their water-borne migration from original deposit sites. The greatest concern lies in predicting the temporal evolution of a contaminant plume under highly heterogeneous soil and rock conditions, and where it is subject to a wide range of geochemical and biochemical transformations. Fractured rocks and karst terrain present particularly difficult challenges in understanding solute transport processes.

Within the upper part of the earth's crust, rocks undergo an important sequence of chemical and physical changes, collectively called weathering, which gradually convert the rocks to soil. Soil lies at the intersection of the two major systems of the external engine, the physical climate system and the biogeochemical cycles. These two systems exchange energy and matter through their interactions, many of which are hydrologically controlled.

Whether adequate soils survive in which to grow crops; whether rivers are navigable; whether there is magnificent scenery: each depends on geomorphic processes driven by water. Much remains to be learned about the processes of erosion and sediment transport, including the effects of varying climate and land use.

Rivers are the conduits for the transport of the water, sediment, and nutrients that control the fertility of floodplains. A quantitative understanding of the mechanisms that will allow the prediction of long-term landscape evolution and the effects of major human interventions is missing. The mechanisms of transport in a river basin are organized around the channel network—a tree-like structure with remarkable properties. How topography differentiates into channels and hillslopes is one of the key questions in its development. What are the unifying principles behind the three-dimensional network geometry? These principles are central to the runoff-generating process, which is intimately linked to the growth and development of the drainage network.

River runoff itself is a key flux in the physical climate system. It is an input to ocean dynamics and an output from the convergence of atmospheric water vapor. This flux highlights the relationship of hydrology and climate. One challenge we still face is to improve our understanding of the interaction between the hydrologic cycle and the general circulation of the ocean-atmosphere system. There is a constant exchange of water between the reservoirs of these two systems, mainly through precipitation, runoff, and evaporation, but the time and space scales of the exchanges vary greatly among the components and with location. Unanswered questions relate to the atmospheric pathways of evaporated moisture and to the sensitivities of atmospheric dynamics to the exchanges of heat and moisture between land and atmosphere. The operational tools in these studies are the atmospheric general circulation models (GCMs) that are being developed to reproduce the basic patterns and processes of atmospheric systems. Only recently have these models been used to study the spatial and temporal patterns in the atmospheric and surface branches of the hydrologic cycle. It is critical to intensify these efforts to quantify the role of land surface-atmosphere feedbacks in the maintenance of climatic systems.

For a GCM to successfully simulate climate and be useful in regional hydrologic studies, realistic modeling of land surface processes is essential. Given that GCM grids are typically 10^4 to 10^5 km^2, the significant effects of spatial heterogeneities in surface hydrologic processes must be defined. Identifying those effects, what controls them, their magnitudes, and their appropriate parameters is among the challenges that lie ahead.

Of all the processes in the hydrologic cycle, precipitation in its various forms has perhaps the greatest impact on everyday life. Atmospheric processes that produce precipitation operate over a variety of space and time scales. They exhibit control and feedback mechanisms, and they interact with surface topography, soil moisture, and vegetation. A characteristic feature of rainfall is its extreme variability over time intervals of minutes to years and in space ranges of a few to thousands of square kilometers. One of the major challenges for hydrologists, meteorologists, and climatologists is to measure, model, and predict the nature of this variability. In hydrology, a primary interest lies in the dynamics governing the time and space distributions of rainfall, especially heavy rainfall that can produce floods, and in understanding the dynamic interaction of the drainage basin with these storms. This requires a link between deterministic models of rainfall dynamics and stochastic models of rainfall fields.

The interaction between land surface processes and regional weather is another exciting frontier in hydrologic science. For instance, under what conditions will the spatial distribution of evaporation generate regional circulations that could influence mesoscale rainfall and regional climate? In this and other questions, it is becoming clear that spatial distribution of the phenomena plays important roles in controlling the strength of the feedback mechanisms between the surface and the atmosphere.

Surficial processes are those involving the transport of mass and energy through the interface between the lower atmosphere and the earth's surface. Once again it is necessary to understand the relevant processes on different temporal and spatial scales. How can local observations of infiltration and soil moisture be translated to larger regions? At the laboratory scale, many important issues remain theoretically unresolved, an example being the effect of the chemical constituents of the soil on its hydraulic properties. At the hillslope scale, debate exists over the role of different factors on the effectiveness of the various flow paths. At the mesoscale, much progress is needed in the formulation of appropriate parameters for regional evaporation, and we need to learn more about which phenomena in the atmospheric boundary layer control evaporation from the land surface and from large water bodies.

The frozen environment presents its own challenges: the behavior of surface and subsurface waters at all scales of description is complicated by phase changes and by the peculiar properties of ice. In alpine terrain, there is a need for methods that integrate the radiation balance over large areas to provide estimates of times and rates of snowmelt.

Surficial processes not only provide key interactions between terrestrial surface moisture and energy and atmospheric dynamics, but also constitute a vital link between the physical climate system and biogeochemical cycles.

The hydrologic cycle provides a useful framework for interpreting key biological processes. From an ecosystem perspective, water is important as a carrier, a cooler, a substrate, and a mechanical force. The dependence of life on water is fundamental since water is the major constituent in essentially all functioning organisms. Their intricate life cycles are organized in most cases around their access to water. Thus the hydrologic cycle represents a fundamental physical template for biological processes.

Hydrology and biology interact over a wide range of spatial scales, from the microscale of small habitats, through the mesoscale of drainage basins, to the macroscale of continents. Similarly, their temporal interactions range over minutes to centuries. **It is precisely in the interplay between the different scales that the hydrologic cycle offers unique scientific challenges in the search for general principles.** In plant dynamics it is known that the abundance of species and their spatial distribution are related to environmental conditions called ecological optima. A natural speculation is that these conditions are the preferred operating domain of the climate-soil-vegetation system. The key question relates to what the optimality criteria are that direct the functioning of this system. Where water is the driving force of the system, a logical hypothesis is that the optimality criteria are related to the key hydrologic variables. Among the exciting scientific challenges are the search for the optimality criteria under different types of constraints and the mathematical representation of these criteria. The fluxes of the key biological variables (e.g., biomass) are also intertwined with the operating chemical processes (e.g., carbon assimilation), which in turn are linked to hydrologic processes (e.g., evapotranspiration). From such relationships it is clear that hydrology is a fundamental structural component of the biogeochemical cycles.

Knowing where a parcel of water has been, and for how long, often is the key to understanding its chemical evolution. Hydrologic residence times are basic unknowns related to many contemporary environmental concerns. Acid rain is a clear example of the importance of understanding and predicting the effects of the chemical composition of precipitation. The use of rainfall composition data as tracers of the hydrologic cycle offers singular opportunities to better understand the relationship between the chemistry of rainfall and the chemistry of soils, ground water, and surface waters. Hydrology and a number of chemical processes also are tightly connected in understanding the effects of soil and vegetation systems on the biogeochemical cycles of nutrients and toxic elements that affect water quality.

Sediment transport, a process long of interest to hydrologists, is receiving much renewed attention as an important vehicle for the storage and movement of chemical species. The nature of organic coatings on stream sediments and the transport rate of polluted aggregates play important roles in the quality of stream waters.

Understanding the interaction of processes at widely different scales is again a pivotal challenge. For example, soil history at a point is largely controlled by microscale chemical kinetics that can be studied in the laboratory, but how is this related to weathering rates at the regional scale?

The biogeochemical cycles of elements like carbon and nitrogen are intimately linked with hydrologic processes. The importance of this linkage cannot be overstated because it affects the very nature of life on the planet and is a major component of the earth's external engine.

A recurrent theme throughout this exploration of the frontiers of hydrologic science is the dynamic nature of the processes involved. These processes are highly nonlinear and have a wealth of feedback mechanisms operating over a wide range of temporal and spatial scales. These features are common to many scientific fields, and their study draws on the same mathematical ideas.

How does one contend with the scale issues that are so pervasive in hydrologic phenomena? Is there hope for unifying relationships across scales? A challenging task is to uncover the organizing structure hidden in the highly irregular patterns that hydrologic phenomena show at different scales. A striking feature of many natural processes is that changes in their scale of description lead to fluctuations that look statistically similar except for a factor of scale. A characteristic irregularity seems to exist that reflects an underlying structure. This characteristic irregularity can be described in terms of what is called the fractal dimension. Recent analyses of spatial rainfall and channel gradients in drainage networks suggest a more subtle type of structure where more than one factor of scale is involved. This is called multiscaling invariance, and it offers an exciting perspective for bringing unifying principles across scales to highly erratic hydrologic phenomena.

The issues of nonlinear dynamics and the limits of predictability are other frontiers of contemporary science intimately related to hydrology. Recent developments in the theory of dynamical systems show that many nonlinear deterministic phenomena are sources of intrinsically generated complex behavior and unpredictability. In fact, they look as if they were a stochastic process, and thus the phenomenon is called deterministic chaos. Is this phenomenon detectable in hydrologic processes? If so, then, there is hope that through its characterization many important features can be understood regarding the complex nonlinear dynamics underlying the processes.

What follows is a more detailed examination of selected frontiers in hydrologic science. **In choosing these topics, the committee has subjectively sought the interesting and exciting, seeking to transmit the flavor of the science rather than to provide either an exhaustive or a rank-ordered list of the most important opportunities.**

HYDROLOGY AND THE EARTH'S CRUST

Introduction

Geoscientists describe the earth in terms of its three major structural zones: the core, the mantle, and the crust. The crust is the rigid outer skin of the earth; in continental areas, it varies in thickness from approximately 15 to 70 km. Two aspects of the upper 10 km of the continental crust that make it unique are that (1) it is the only region of the earth's subsurface to which humans have direct access, and (2) it forms the interface between the earth's two major dynamic systems—the hydrologic system and the tectonic system. The tectonic system—and the concept of plate tectonics—involves a grouping of processes that lead to the formation and deformation of crustal rocks. Whereas the hydrologic system is set in motion primarily by solar energy, the tectonic system is driven by the earth's own internal thermal energy. When the role of hydrology in tectonic processes is considered, the depth scale of interest is kilometers, with horizontal distances usually on the order of tens to hundreds of kilometers. The processes of interest are the movement of fluids and the transport of mass and energy in the earth's crust.

Hot springs provide an example to illustrate the nature of the interaction between the tectonic and the hydrologic systems. Hot springs develop when meteoric water originating from rainfall or snowmelt circulates to depths of several kilometers, adsorbs heat from the surrounding rock matrix, and then is able to move relatively quickly to the ground surface along fault zones. Figure 3.1 shows how heat flow from deeper levels of the earth's crust can be captured by the ground water flow system and diverted to a major fault zone. In areas where subsurface temperatures are increased by local intrusions of molten

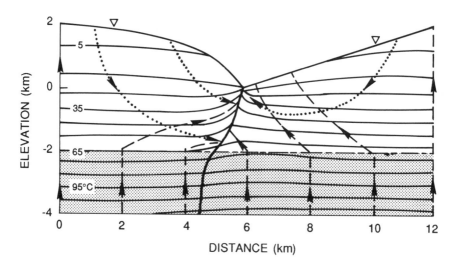

FIGURE 3.1 Patterns of fluid flow (dotted lines) and heat transfer (dashed lines) in an asymmetric mountain valley. This diagram shows how ground water flow can modify the subsurface thermal regime, by adsorbing heat and transferring it to a fault zone (thick line). A warm spring discharges in the valley. SOURCE: Reprinted, by permission, from Forster and Smith (1989). Copyright © 1989 by the American Geophysical Union.

rock, hydrothermal systems may develop. The hot springs and geysers of Yellowstone National Park are dramatic examples. Active hydrothermal systems are potential sources of geothermal energy. Relic geothermal systems are targets for mineral exploration because metals may have been transported by the hot fluids, and geochemical conditions may have promoted precipitation of those metals to form ore deposits.

Different issues arise when we focus on processes occurring within the upper several hundred meters of the earth's surface. This vadose zone normally comprises the weathered, unconsolidated soil material that is present at the land surface. In the vadose zone, both air and water are present within the open pore spaces between the solid grains. The medium is the site of innumerable chemical transformations mediated by solar radiation, wet and dry atmospheric deposition, and biologic activity. The vadose zone is a storage component of the hydrologic cycle, a reservoir of water, air, and reactive inorganic and organic solid matter. It influences the runoff cycle and ground water recharge by affecting both the flow patterns and the quality of surface and percolating subsurface waters.

The water table marks the transition from the vadose zone to the deeper saturated ground water zone, where all the pore spaces are filled by water. Like the vadose zone, the saturated zone is a reservoir of water and supports a range of chemical reactions. Issues at this scale center on the characterization of the physical, chemical, and biologic processes occurring within the subsurface hydrologic environment, their link to hydrologic processes occurring on the earth's surface, and the development of techniques for quantifying these processes and monitoring their effects. Many research questions here have direct relevance to the serious environmental problems facing our society.

A number of processes must be addressed at the microscale, that is, the scale of the individual pore spaces within a soil or rock. The transport of chemical species dissolved in the water is a complex and dynamic process. Solutes entering the subsurface can interact with other dissolved solutes, with the solid matrix, and with the native ground water and can take part in the life cycle of microbes present within the subsurface. **There is feedback between these biochemical processes and the patterns and rates of fluid flow. Greater understanding at the microscale is necessary to build a framework for developing predictive models that apply at the mesoscale, the scale at which it is feasible to tackle most applied problems in subsurface hydrology.**

Some Frontier Topics

The Role of Ground Water in Tectonic Processes

To what extent can the application of quantitative hydrogeologic concepts provide new insights into geologic processes that occur in the earth's upper crust?

Water originating as precipitation is thought to be able to penetrate to depths of at least 10 km. Water present within the void spaces of sediment or rock is a central feature in a number of geologic processes because (1) fluid pressures influence the strength of sediments and rocks to resist shearing and thus influence processes such as landslides, faulting, and earthquakes, and (2) fluid flow is the key process for large-scale redistribution of mass and heat within crustal rocks. Although rates of ground water flow are much lower than those in the upper few hundreds of meters of the earth's surface, and time scales may approach 10^6 years or longer, from a geologic viewpoint ground water circulation within the upper crust is no less important than the near-surface component of the hydrologic cycle.

Permeability is the parameter that quantifies the ability of a fluid to flow through the interconnected pore spaces of a rock or soil. Figure 3.2 identifies typical values of permeability for a variety of geologic deposits and rock types. Hydraulic conductivity, the permeability when the fluid is water, often is used to characterize the flow of water through near-surface soils or rocks. More permeable sediments or rocks, capable of transmitting significant quantities of water, are referred to as aquifers. The wide range of variation in permeability implies that subsurface fluid fluxes (flow per unit area) can vary by orders of magnitude, depending on the nature of the geologic setting.

Recognition of the significant role of circulating fluids in tectonic environments is not a recent development. The geologic literature contains a vast array of models that have been proposed to explain innumerable sets of data. However, for many years the science proceeded no further than well-reasoned qualitative analysis. To quantify this link between the hydrologic and tectonic systems, the nonlinear interaction of the hydraulic, geochemical, stress, and thermal regimes must be tackled. Stresses in the crust originate from movements of the earth's tectonic plates, and more locally, from the weight of overlying rock units. A benchmark paper by M. King Hubbert and William

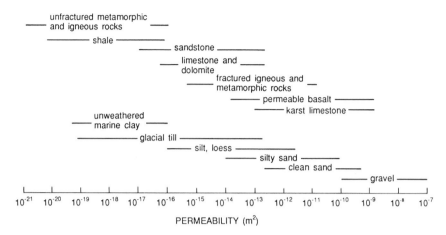

FIGURE 3.2 Permeability of common geologic media. SOURCE: Adapted, by permission, from Freeze and Cherry (1979). Copyright © 1979 by Prentice Hall, Inc.

Rubey demonstrated the importance of pore fluid pressures in the mechanics of faulting, and ultimately, in mountain-building processes (Hubbert and Rubey, 1959). This work, probably more than any other, set the stage for interactions between ground water hydrologists, geologists, and solid-earth geophysicists. In the early 1970s, Barry Raleigh, Jack Healy, and John Bredehoeft, in what have become known as the Rangely experiments, provided field confirmation that fluid pressure can be a key parameter in triggering earthquakes associated with faulting (Raleigh et al., 1972). Others documented the fact that filling a large reservoir behind a new dam can induce local earthquake activity. Dennis Norton and his colleagues at the University of Arizona were among the first to promote a quantitative framework to link geochemical processes, fluid circulation patterns, and heat transfer (Norton, 1984). Recently, attempts have been made to quantify the role of ground water flow in regional metamorphism, where, for example, a rock such as limestone is transformed to marble. The concepts and tools of hydrologic analysis are being adopted to solve a number of fundamental geoscience problems, and opportunities abound for collaborative research.

Earthquake Cycle Earthquakes occur when slip is initiated along a fault and stored energy arising from long-term tectonic movement is abruptly released. Subsurface waters, in responding to changing pressure, thermal, and stress conditions, can have a significant impact on the

M. KING HUBBERT
(1903-1989)

The word "geophysics" is much broader that the narrow application of geophysical tools to mineral and petroleum exploration. It implies an integration of mathematics, chemistry, physics, and geology in the study of the earth. In this context, it could be argued that M. King Hubbert is the father of geophysics. In his education, his career, his research, his essays, and his speeches, this outspoken maverick led the earth sciences kicking and screaming from a largely observational and descriptive style to a more quantitative, experimental, predictive science.

Hubbert was born in 1903, attended a one-room Texas schoolhouse, received an unconventional high school education, and graduated from the University of Chicago in 1926, having taken a self-designed academic program that emphasized physics, mathematics, and geology. He put little store in degrees and received his Ph.D. only in 1937, on the basis of his publication of *The Theory of Scale Models As Applied to the Study of Geologic Structures*, later revised into his classic paper "The Strength of the Earth." His career spanned academia, government, and industry. He taught at Columbia University in the 1930s and at Stanford University in the 1960s. From 1943 to 1964 he was a research scientist with Shell Development in Houston, and from 1964 to 1976 a research geophysicist with the U.S. Geological Survey in Washington, D.C.

Hubbert made three major contributions to the earth sciences and one to society, and any one of them would have guaranteed him a good measure of fame. His first recognition came from his publication in 1940 of the *Theory of Groundwater Motion*, a treatise that elegantly clarified the fundamental physics that underlies the flow of fluids through porous media. In 1953, he applied these principles in the "Entrapment of Petroleum Under Hydrodynamic Conditions," a paper that eschewed conventional wisdom and eventually had a major impact on petroleum exploration strategies. In 1959, he and William W. Rubey attacked what was then one of the most enigmatic paradoxes in geology, the process of thrust faulting. Their classic paper, the "Role of Fluid Pressure in the Mechanics of Overthrust Faulting," introduced the concept of effective stress in a structural geology framework. It opened the door to later work by others on earthquake prediction and control.

Hubbert's societal contribution revolves around his claim, first made in 1949, that the fossil fuel era of energy production would be relatively short lived. His bell-shaped curve (often called the "Hubbert Pimple"), which traces the complete cycle of U.S. and world crude oil production from discovery to exhaustion, first alerted the public to the fact that petroleum resources are not inexhaustible. Although hotly contested over the years, his projections have had a major impact on the international oil industry and on U.S. government policy.

Hubbert's papers are models of elegant clarity. He taught three generations of earth scientists, by his example, to think clearly, to write clearly, and to fiercely question accepted geological wisdom if it does not conform to fundamental physical and chemical principles.

earthquake cycle. Three aspects of earthquake mechanics are relevant here: (1) the generation of high fluid pressures within fault zones, (2) expansion in the presence of pore fluids, and (3) pore pressure fluctuations prior to and during earthquake events. Expansion refers to the change of pore volume in the presence of an applied differential stress. Lithostatic pressure at a particular depth is equal to the total weight of the overlying rock and water. As fluid pressures approach lithostatic pressures, rocks lose their strength to resist a shear force.

When rocks slide past each other along a fault, friction generates heat. Thermal pressurization within a fault zone during faulting can cause fluid pressures to rise to near-lithostatic values, leading to strain weakening as the earthquake progresses. Richard Sibson (Sibson, 1973), and Art Lachenbruch and John Sass (Lachenbruch and Sass, 1980), carried out early studies of the process. More recent analyses using computer models have shown how this behavior can influence the character of fault motion, including earthquake magnitude. The hydraulic characteristics of the fault zone and the surrounding rock are key parameters in this process.

Currently, there is only limited observational evidence to confirm the results of theoretical models. Limited knowledge of the hydrologic and deformation properties of fault zone materials restricts progress in advancing the theoretical framework to explain the mechanics of shallow earthquakes (depths less than 15 km). Such information is basic to the development of a capability for earthquake prediction. Spatial variations in hydraulic properties along the trace of the fault may be important in explaining the size and magnitude of barriers that act to resist fault motion and influence the magnitude of slip during single events. Research is needed to identify methods by which these data can be obtained.

As tectonic stresses accumulate in the rock adjacent to a fault, changes in fluid pressures cause water level fluctuations in wells and affect the strength characteristics of the rock. Detailed observations of water level fluctuations that can be clearly linked to the earthquake cycle may be available within several years. The hope is that measurements of this kind may eventually help us predict the timing and location of earthquakes. Volume recovery due to earthquake-induced stress drops may cause a pore pressure rise and act as a triggering mechanism for further earthquakes. Improvements in our understanding of field-scale hydrologic properties of rocks under changing stress conditions, and methods for obtaining those data, are key research needs. Intervention in the earthquake cycle to "manage" earthquake hazards remains a distant goal.

Sedimentary Basins Sedimentary basins form a substantial portion of the land mass of North America. They originate as regions of the continent where the crust was downwarped in the geologic past and infilled by sediments. These sediments eventually are transformed to sedimentary rocks. Near-surface regions of sedimentary basins, to depths of roughly 500 m, have been developed extensively to meet water supply demands. Other resources occur at greater depths. All of the continent's oil and gas reservoirs are located within sedimentary basins. Ground water flow plays a role in the transport of oil from source rocks to entrapment in a reservoir. Because ground water flow can modify the thermal regime of a basin, it is also a key consideration in the maturation of organic matter to produce oil. Fluid pressures approaching lithostatic values can develop during the evolution of a sedimentary basin because of the combined effects of compaction of the sediments, heating of confined fluids, and dehydration of clay minerals. These overpressures modify the flow patterns from those that would be predicted solely on the basis of ground water recharge from regions of higher elevation within the basin. The presence of overpressured zones also increases the hazards and costs associated with drilling operations to find new oil and gas reservoirs.

Hydrologic models are being developed to investigate physical, chemical, and thermal relationships among ground water flow, transformation of sediments to form rocks, hydrocarbon accumulation, and evolution of the basin through geologic time. The availability of supercomputers has allowed researchers to investigate the nature of these coupled processes on large spatial and temporal scales in ways not possible only 10 years ago. However, important gaps remain in our understanding of the mechanisms involved. For instance, the deformation of porous media is a time-dependent process. For short-duration events with time scales up to 10 or 100 years, elastic models are adequate to describe deformation. In an elastic model, deformation is reversible (examples of this include the stretching of a rubber band or compression of a steel spring). Biot (1941) presented an elegant theory to describe the elastic deformation of porous media. The coefficients that describe the elastic deformation can be measured in laboratory experiments or from tests in boreholes. However, for processes that operate on a geologic time scale, such as sediment deposition and basin subsidence, elastic models appear inappropriate. Long-term deformation is dominated by physical and chemical changes that are irreversible. Inelastic coefficients are thought to be of significantly greater magnitude than short-term elastic coefficients. Better understanding of inelastic coefficients, and the development of methods for estimating their values, constitute an important area of research toward improving our capability to simulate patterns of ground water flow on a geologic time scale.

Discharge areas for mesoscale ground water flow sometimes are associated with important mineral deposits of metals, such as lead and zinc. Metals released from source rocks are transported to sites where geochemical conditions promote the formation of sulfide minerals. Over geologic time, large ore deposits can accumulate. Such a situation is thought to explain the lead-zinc deposits of the upper Mississippi Valley. If it were possible to reconstruct the hydrogeologic and geochemical evolution of a basin, including transients imposed by erosion, tectonic stress, and thermal stress, then better exploration methods might be developed. Recent research has formulated the theoretical basis for carrying out these kinds of simulations. However, theories on transient flow are difficult to test with observational data because of the long time scales involved. A key research goal is the development of methods for building a sufficiently robust data set to constrain models of flow patterns that existed in the geologic past.

Large-Scale Flow and Transport Processes The greatest obstacle to progress in understanding the role of pore fluids in tectonic processes is the lack of in situ data on the hydrologic properties of crustal rocks and fluid pressures, together with the lack of knowledge of their spatial and temporal variability. A new quantitative approach that builds on sound hydrogeologic principles and techniques is fostering great strides in our understanding of the complex nature of tectonic processes within the upper crust, but experimental confirmation is needed. Research teams of geologists, geophysicists, and hydrologists are being formed at institutions around the world. Hydrologists are contributing a realism to the models of fluid flow that was lacking in many earlier studies.

Progress is limited, however, by data deficiencies. For the great variety of geologic environments that constitute the continental crust, insufficient data exist on the magnitude and variation of pore pressures, fluid properties, and the nature of porosity and permeability. This creates fundamental problems in identifying realistic values for medium properties to use in model calculations. Most studies are based on the concept of homogeneous rock properties, but there is little reason to expect that spatial variability will be any less important at depth than it is for near-surface environments. Similarly, there is a paucity of detailed data to test hypotheses and model assumptions rigorously. The National Science Foundation's Continental Scientific Drilling Program is one coordinated effort to provide data of the kind needed to guide scientific inquiry.

RADIOACTIVE WASTE DISPOSAL

Highly radioactive waste materials generated during the production of nuclear power are currently held in temporary storage at nuclear power plants across the United States. If nuclear power is to remain viable, long-term plans to deal with nuclear waste must be set in place. The Nuclear Waste Policy Act of 1982, as amended in 1987, directs that the U.S. Department of Energy develop a program for permanent disposal of nuclear waste in an underground geologic repository. The plan is to build a repository that would be much like a conventional mine, with the waste materials emplaced in containers and then set in holes along the walls or floor of the repository. After receiving waste for a period of 30 to 50 years, the underground openings would be backfilled and the site closed.

Radioactive waste is hazardous for time periods that exceed tens of thousands of years. The concept of geologic disposal is founded on the idea of multiple barriers: first, an engineered system to encapsulate the radioactive waste within the repository and, second, the natural system, the surrounding rock formations. It must be anticipated that the engineered barriers will eventually fail and that the radionuclides will begin a slow journey back to the ground surface, carried along by naturally occurring ground water flow in the host rocks. This ground water travel time from the repository to the accessible environment is a key factor in determining the suitability of a particular location for the repository.

Current performance criteria set down by U.S. regulatory agencies include specifications related to radionuclide arrival rates at the earth's surface and the pre-emplacement ground water travel time from the region of the repository to the accessible environment. The pre-emplacement ground water travel time along the fastest path of likely radionuclide travel must be at least 1,000 years. Our ability to predict the ground water travel time, and the concentrations of radionuclides in the ground water, will determine, in large part, the environmental risk that future generations will face if this disposal option is adopted. Given the length of time involved, observations of present-day conditions do not tell the complete story. Predictions must be based on sound physical and chemical understanding.

A site at Yucca Mountain in southern Nevada is being studied to assess its feasibility as a repository. Because of the desert climate, ground water recharge is limited and the water table is very deep. The repository would be located within a volcanic rock unit above the water table, but still some 300 m below the ground surface. Under current conditions, rates of ground water flow are thought to be on the order of a few meters per century. One of the concerns that must be addressed is how these velocities might change at some point in the future if the climate of southern Nevada were to become wetter.

A major research program has been funded to address the scientific and technical issues associated with predicting how ground water and dissolved radionuclides may move through the subsurface environment at Yucca Mountain. The problem is complex. A broad range of geologic, hydrologic, and geochemical questions must be answered to evaluate whether or not the site is capable of isolating radioactive waste. The geologic repository presents hydrologists with a task never before faced: predicting, with reasonable assurance, the occurrence and movement of ground water flow and radionuclide concentrations on a time scale that extends thousands of years into the future.

Even in a favorable funding climate, it is unreasonable to expect that more than several deep boreholes will be available at any one site. Each borehole yields data representative of conditions a short distance into the surrounding rock mass. It is not yet known to what extent these (already expensive) measurements carry information helpful in quantifying larger-scale flow and transport processes. Methods for integrating these measurements with higher-density geophysical data must be explored if we hope to adequately characterize the subsurface environment.

To what extent do small-scale measurements provide information about large-scale flow and transport processes? How can we relate these measurements to other large-scale geophysical data?

The pore space between the grains of a rock such as sandstone, or the interconnected microcracks within crystalline rocks such as granite, generally decreases with depth. As a result, it is probable that large-scale patterns and rates of fluid flow and mass transfer at greater depths within the crust are controlled by fractures and shear zones. It has yet to be established that conventional modeling approaches, based on assigning medium properties such as permeability to representative volumes of the rock mass, provide the correct framework to quantify mesoscale ground water flow and solute transport in deeper crustal rocks. Furthermore, in active tectonic environments existing fractures and shears will repeatedly open and close, and new fractures will appear. There is much speculation on the cyclic fluctuation of fluid pressures between lithostatic and near-hydrostatic values, with pressure buildup and consequent release associated with fault rupture and locally enhanced permeability. The framework within which to treat the hydraulic properties of rock as transient variables is largely unexplored.

Perhaps the greatest challenge comes in dealing with hydrologic processes that operate on time scales of 10^3 to 10^6 years. Here we are limited to observing the outcome of nature's past experiments. Analyses based on steady state assumptions can only be crude approximations of reality. Although the geologic record contains a wealth of information, we do not yet have the knowledge base to extract the quantitative measures that are needed for experimental validations.

Fractured Geologic Media

What are the links between geologic models of fracture formation, geochemical models for mineral infilling and dissolution, mechanical models of in situ stress conditions, and the statistical descriptions of network geometry used in stochastic fluid flow and solute transport models?

The hydrology of fractured rock is important for several reasons. In many bedrock aquifers, fractures provide an appreciable contribution to the capacity of the medium to transmit a fluid. At greater depths, they may account for the primary contribution, with a much smaller fluid flux occurring within the rock matrix. Because geologic units with low permeability are targeted as preferred sites for land-based disposal of toxic and radioactive waste, fractures represent the most likely pathways for off-site migration of contaminants. The scientific basis for describing fluid flow and solute transport in fractured media lags behind the state of knowledge for describing these processes in porous media systems.

There are two approaches to the description of ground water flow in geologic media containing fractures: (1) the continuum approach, which treats the fractured medium as if it were equivalent to a porous medium, and (2) network models based on an explicit representation of the fractures. Figure 3.3 is an example of a computer-generated model of a fracture system; it shows a three-dimensional network of disc-shaped, orthogonal fractures. The rock matrix between fractures is not shown in this diagram. If fracture density is high, then the

FIGURE 3.3 An example of a computer model of a fracture network. Fractures are represented as disc-shaped features of variable size. SOURCE: Reprinted, by permission, from Long et al. (1985). Copyright © 1985 by the American Geophysical Union.

continuum approach is appropriate. The last decade has seen important advances in the continuum representation of fluid flow in fractured rock. The transport problem is more difficult. Advection and dispersion are understood only in general terms. In the early 1970s, de Jong developed a continuum model for solute transfer in networks of fractures with uniform spacing and infinite length, but a corresponding theory does not exist for the realistic case of finite-length fractures, irregularly located within a rock mass (de Jong and Way, 1972). Even less is known of solute transport processes in unsaturated fractured rock.

In media with relatively sparse fractures, or those where a significant proportion of the fractures are partially or fully sealed by mineral precipitates, transport for a considerable distance may be required before solutes encounter a representative sampling of fluid velocities. Until that distance is reached, continuum models are not applicable. It is also possible that several scales of fracturing occur within the rock mass, with a small number of areally extensive fractures exerting a predominant influence on patterns and rates of fluid flow. In either of these situations, discrete network models are appropriate. Stochastic concepts underlie the development of this approach, with probability distributions defining fracture locations, dimensions, orientations, and apertures. The suitability of models of fracture geometry that are based on concepts of statistical homogeneity has yet to be evaluated critically. Field-based hydrologists must take a lead role here.

If existing approaches for modeling fluid flow and solute transport in fractured rock are of questionable reliability, what new methodologies may be better suited for application in this important geologic setting?

Uncertainties exist in the hydrology of fractured rock at both the microscale and the mesoscale. Fluid flow laws have yet to be determined for rough-walled fractures, with basic questions to be resolved on the nature of fluid pathways in single fractures and on the dependence of fluid flux on the stress field within the rock mass. Little quantitative work has been carried out on the transport of reactive solutes in fracture networks. Geologic studies indicate a close relationship between the hydraulics of flow and the geochemical environment. Existing models of mesoscale solute transport, and their application in practice, are of questionable reliability. Conventional continuum models may not be applicable, especially in sparsely fractured rock or in media with multiple scales of fracturing. The existing alternative, based on a discrete representation of network geometry, is limited in practical application by computational constraints, even when supercomputers are used. It may be that distributed parameter models, based on an assumed form of the differential equation representing fluid flux and transport processes within a representative volume of the rock mass, are not suited to field-scale prediction. Alternative modeling approaches should be investigated.

What are the measurement techniques needed to characterize field-scale hydraulic and transport properties of lower-permeability rocks, including fractured rock masses?

The development of in situ techniques to characterize the properties of fractured media has been, and will continue to be, an important research field. An integrated approach is evolving wherein a broad range of techniques is used to identify the geometric and hydraulic properties of fracture networks. Options include direct methods based on hydraulic and tracer tests and remote sensing methods using seismic, radar, or electromagnetic tomographic techniques to detect, map, and assess the properties of dominant fractures within the rock mass. We cannot yet accurately measure parameters that characterize the fluid flux and transport properties of individual fractures. The next five years should see important advances in the methods for making field measurements.

Spatial Variability and Stochastic Simulation

What are the relationships between the spatial structure of medium and flow system variables, such as hydraulic conductivity and fluid velocities, and the geologic processes forming soils, unconsolidated sediments, and rock units?

Spatial Variability and Geostatistics One of the key advances in subsurface hydrology over the past decade has been the incorporation of the spatial variability of the hydraulic properties of porous media into our theories of fluid flow and solute transfer. Textural and structural variabilities in soils and geologic media are the principal causes of spatial variability. Transport phenomena also exhibit spatial and temporal variability because of sporadic inputs of precipitation, irrigation, and contaminants. Pioneering studies of spatial variability in water and solute transport within the vadose zone were performed by Donald Nielsen, James Biggar, and co-workers at the University of California, Davis (Biggar and Nielsen, 1976). The corresponding pioneering work in the saturated zone was done by R. A. Freeze (1975; Canada), L. W. Gelhar (1976; United States), and J. P. Delhomme (1979; France). These early studies clearly showed the extremely variable nature of water and solute transport parameters (e.g., hydraulic conductivity and velocity of solutes) as compared to the more modest variability of water retention parameters (e.g., degree of saturation and bulk density). These trends have been confirmed in many subsequent field experiments (Table 3.1).

Characterization of spatial variability entails field sampling and statistical analysis to infer means, standard deviations, and the length scale over which correlations are significant. This so-called correlation length is a measure of distance over which neighboring values are likely to be of similar magnitude. These ostensibly simple operations involve challenging problems, both physical and mathematical. Subsurface hydrologic data are collected from a limited number of sites, each of which is used to sense, with some degree of error, conditions in a small volume around the measurement point. Adding to this

TABLE 3.1 Spatial Variability of Water and Solute
Transport Parameters Observed in Field Studies

Parameter	Coefficient of Variation (percent)[a]
Porosity	6-12
Water content	4-45
Bulk density	3-26
Hydraulic conductivity	48-320
Infiltration rate	23-97
Solute velocity	61-204

[a]The coefficient of variation equals 100 × standard deviation/mean
for samples collected over a field-scale region.

SOURCE: Copyright © 1985. Electric Power Research Institute.
EPRI EA-4228. "Spatial Variability of Soil Physical Parameters
in Solute Migration: A Critical Literature Review." Reprinted
with permission.

observational problem is the prohibitively large number of samples
required, in principle, to characterize a parameter that varies by several orders of magnitude.

Statistical methods to characterize spatial variability have been applied in a number of hydrogeologic settings. Recent studies have
demonstrated the extreme sensitivity of the estimate of the correlation
length to the precise nature of the procedures used in estimating its
value. Failure to subject a data set to a comprehensive analysis may
result in estimates of correlation length values that have little physical meaning. Other issues that warrant further study include (1) the
identification of regions with similar medium properties and possible
multiple scales of heterogeneity, (2) the incorporation of interpretive
geologic models to augment borehole measurements, and (3) an assessment of the viability of statistical methods when only sparse data
sets are available or can be collected. This latter situation is frequently
the case in studies of deeper ground water flow systems.

Measurements of fluid pressure and the collection of representative water samples can present significant technical challenges, especially
in deeper boreholes and in lower-permeability systems. Recent improvements in instrumentation, such as modular multiport samplers
and borehole velocity meters, have expanded the capability to obtain
more detailed and accurate data from individual sites. New computational techniques are required to process the higher-density data
sets available from multiport samplers.

An important research opportunity lies in the development of general relationships linking the processes forming a soil or geologic deposit, and the resulting spatial structure in transport properties for water and solutes. Observations are required from a wide range of soil and geologic environments throughout the mesoscale (1 to 100 km) range. The same kinds of improvements are needed for field observations of solute concentrations, with the added complexity that high-quality data varying over time scales of months and years are required. Improved data bases are needed to examine the critical issue of whether or not the statistical properties of a porous medium are spatially uniform or if they change with location. Virtually every model of fluid flow and solute transport in heterogeneous media is predicated on the assumption of spatially uniform statistical properties.

Current stochastic theories of transport predict only ensemble averages. What criteria should we use to establish the equivalence between the transport response in a field experiment and the ensemble average?

Stochastic Analysis and Prediction Uncertainty Solutes that enter a ground water flow system are carried away from the point of entry in the direction of ground water flow. This process is known as advection. As the "plume" of solutes moves farther from its source location, it continually expands in size, in much the same way as a plume of smoke that is carried away from a smokestack by air currents. This spreading process is known as dispersion. Because of the many heterogeneities within porous media, processes should be modeled mathematically as random functions of space and time, that is, as stochastic processes. A stochastic process consists of an ensemble of realizations, which is the set of all possible numerical values of the process. The common assumption in stochastic modeling has been that the dispersive and advective properties of a mesoscale soil or aquifer are single realizations of random functions, based in turn on an assumed stochastic model for hydraulic conductivity or solute velocity. This form of the transport equation is termed a "stochastic advection-dispersion equation." Stochastic models based on this approach predict the ensemble-average solute concentrations.

In modeling solute transport in the subsurface, the main interest is not with the ensemble behavior, unless its relevance to prediction for a single field soil or aquifer can be established. There is a need to develop rigorous methods to connect the predictions of an ensemble-based stochastic model to experimental observations made in a field study. A single field experiment measures a single realization of a stochastic process, not its ensemble average, but nearly all current theories predict only ensemble averages. Studies of the variance of a random function under conditions found in subsurface environments are needed to establish the criteria for equating a measured transport property with its ensemble average.

What kinds of field experiments are needed to assess the calibrations and predictive application of stochastic transport models at length scales of 100 to 1,000 m?

The current use of stochastic concepts represents an initial attempt to capture field-scale variability in terms of a set of physical and mathematical approximations. However, many of the approximations remain both untested in detailed field experiments and underived from rigorous results in the theory of random processes. Calibration and the predictive application of stochastic models make demands on experimental data bases that no field study to date has met satisfactorily. One requirement that is self-evident but surprisingly difficult to fulfill is mass conservation: the field sampling procedures and data interpolation methods should lead to a calculated total mass in the subsurface zone that agrees with the mass known to have entered the zone. Another important requirement that is not typically met in field studies of transport is specification of the initial conditions. If a stochastic model is to simulate the evolution of a solute plume, for example, it must have as input a precise numerical account of the solute distribution at some initial time. Stochastic transport theory has yet to be evaluated critically in settings where the transport distance is on the order of 100 to 1,000 m, the scale of greatest concern at most hazardous waste sites. These experiments will be time consuming, tedious, and expensive, but of great scientific value.

What are the theoretical approaches and experimental methods appropriate to aid in the identification of model structure and parameter estimation for simulating subsurface fluid flow?

Before a model can be used for prediction, it must be calibrated. In 1935, C. V. Theis set the stage for estimating aquifer parameters by first reporting how ground water flow to a pumping well can be described as a mathematical boundary value problem (Theis, 1935). Given this formulation, it is possible to estimate medium parameters such as hydraulic conductivity by measuring drawdown in one or more observation wells. Since that time, innumerable analytical techniques have been developed to characterize more complex hydrogeologic settings. With the introduction of numerical computer models, calibration took on a different format. Today, methods that can be collectively titled as "inverse simulation" are at the research frontier. These methods are closely linked to stochastic simulation. In most instances, there is considerable uncertainty in the values of model parameters and in the assignment of boundary conditions. This problem is compounded when measurement error is considered. Continued research is needed to aid in the identification and validation of model structures, in the determination of their scale dependence, and in the estimation of parameter values—incorporating both direct and indirect measurement, geologic models, and hydrologic judgments. Progress here will ultimately improve efforts to develop a more precise understanding and assessment of valuable aquifer systems.

Dealing with the Complexity of Reactive Solutes

Although there has been growing appreciation of the effect of the "unmappable" heterogeneity of water-bearing strata, much of our understanding of basic subsurface processes depends on our still rather imperfect capability to quantitatively analyze solute transport under homogeneous and piecewise-homogeneous (i.e., "mappably" heterogeneous) circumstances. To understand the reasons for this still-limited capability, consider the complexity of the situation. Imagine a small volume about some point within the vadose or saturated zone, which is initially in physical and chemical equilibrium. Suppose some new solution invades this pore space, perturbing the original equilibria and throwing the system into a rapid state of change. The new and original solutions mix in a complex way that is determined by the pore geometry. Their solutes react with each other, forming new compounds. Many of these solutes also react with substances encountered on the mineral surfaces or in the gas phase. Some of the original solids slowly dissolve. Newly formed solids precipitate on certain mineral surfaces, but some of the minute particles are carried by water for limited distances before getting stranded in the pore-space maze. A similar fate awaits colloidal particles formed by chemical reactions. These processes change the fluid-conducting properties of the subsurface. In turn, these changes influence the magnitudes and

C. V. THEIS
(1900-1987)

A research scientist's fame may rest on a lifetime of small stepwise achievements, or it may rest on a single contribution of lasting importance. C. V. Theis ensured his fame with a single five-page paper that appeared in 1935 in the *Transactions of the American Geophysical Union.* Entitled "The Relation Between the Lowering of the Piezometric Surface and the Rate and Duration of Discharge of a Well Using Ground Water Storage," it earned Theis recognition as the founder of the applied science of transient ground water flow. Its date of publication is often taken as the birthdate for the application of modern quantitative methods of field investigation in hydrology. Theis's contribution is not likely to be forgotten; in most hydrogeology textbooks, only two headings reflect the name of their originators: Darcy's law and the Theis curve.

Theis was born in Newport, Kentucky, in 1900. He graduated from high school in 1916 and from the University of Cincinnati in civil engineering in 1922. In 1929, after serving as both instructor and student in the geology department at the University of Cincinnati, he received the first Ph.D. ever awarded by that department.

His career was spent almost entirely with the U.S. Geological Survey, first in Washington, D.C., and then in Albuquerque, New Mexico. He was studying the ground water resources of the Portales Valley of New Mexico for the USGS when he began to think about an in situ method of measuring aquifer permeability based on recording piezometric drawdowns in an observation well near a pumping well. His method is based on a mathematical solution to the appropriate boundary value problem that he developed by analogy with the available heat flow solutions. In this exercise he was helped by his friend, C. I. Lubin, of the University of Cincinnati, who is acknowledged in the paper but declined co-authorship. Theis's 1935 paper inspired a lengthy succession of aquifer hydraulics research in the 1940s, 1950s, and 1960s by others. In the 1950s, Theis became the USGS liaison with the Atomic Energy Commission, and he wrote many early articles on the hydrogeologic issues associated with nuclear waste disposal.

Colleagues who knew Theis at the USGS recall him as a hard-nosed, fiesty leader of the highest personal integrity, a man who admired clear thinking and who always had time for younger colleagues.

distributions of solute fluxes and concentrations. Some of the incoming solutes may activate dormant spores, which germinate and become microbes, which then absorb and transform some of the solutes passing near them, creating a variety of new substances.

No doubt, a considerable intellectual challenge is presented by the problem of understanding such a tangled, dynamic, multivariable system. Yet such an understanding is necessary to analyze the processes that determine the chemical evolution of natural or human-influenced waters. Early methods to deal with the transport of reactive solutes often were adopted from the chemical engineering literature. However, significant differences exist between solute transport processes taking place in fixed-bed chemical reactors and those occurring in sediments of aquifers or soils. Methods for describing solute transport in geologic media have had to outgrow their roots in chemical engineering and other nonhydrologic sciences. This state has now been reached, calling for special efforts and opening interesting possibilities.

The physics of solute transport in homogeneous porous media is now fairly well known. The main exceptions involve transport in porous media containing both air and water (or other immiscible fluids). This process is poorly understood and requires much attention. For instance, unexplained and contradictory data exist concerning the dependence of solute-mixing processes on volume proportions between air and water within the pore space. A frequently used assumption that may require examination is that, in porous media, only dissolved substances are mobile. Mobility of small particles and microorganisms has reemerged as an important topic of study, involving surface chemistry, biology, and physics.

What are the kinetics of geochemical reactions involving more than one phase, and what model should be used to account for the complex geometry of the solid-fluid interface that exists in geologic media?

The transport of reacting solutes involves still imperfectly understood geochemical kinetics. Particularly insufficient is our understanding of extremely important reaction kinetics that involve more than one phase (e.g., one involving a mineral solid and a water-carried solute). These kinetics need to be studied in situ so that the influence of the subsurface pore geometry can be taken into account. In the case of heterogeneous kinetics, transport analyses are generally based on the so-called pseudohomogeneous approach or approaches developed in connection with chemical reactor problems. The pseudohomogeneous approach assumes that the location of the reaction is within the body of the solution rather than at the solid-solution interface. It is very important to study experimentally and theoretically the correctness and limitations of such a basic assumption. The second type of approach is a reasonable one for a reactor with man-made components. In the subsurface the pore geometry may be very different. The influence of "geologic" pore geometries is yet to be fully understood.

How can we quantify microbiological processes that occur in the vadose and saturated zones, and the interactions with the physics and chemistry of subsurface transport?

Consideration of microbiological processes in the integrated analysis of the physics and chemistry of subsurface transport, especially the deeper subsurface, is a relatively recent phenomenon. It adds another level of complexity that in all probability will be as large and as important as the chemistry level. Many of the mathematical expressions used for the relevant process rates involve an analogy to enzyme kinetics. The completeness of this analogy should be the subject of more critical studies. The entire field of quantifying microbial effects is in its initial stages of development. The theory of biological reactors, which assumes that an excess of nutrients exists, is still very influential. Microbial ecology under field conditions of severely limited mineral nutrients and energy-yielding compounds has yet to be critically studied.

Contamination of Ground Water Flow Systems

How should we quantify the processes that determine the transport and fate of synthetic organic chemicals that enter the ground water system?

A major environmental focus of the past decade has been the recognition of the extent to which ground water has been, or potentially is, at risk of being contaminated by inorganic and synthetic organic chemicals. Many examples are available where industrial waste management practices, agricultural production, and natural resource developments have led to the introduction of hazardous chemicals into a ground water system. Contaminants found in ground water are of three general types: (1) chemical species that are miscible with the pore water and migrate with the ground water flow as a dissolved aqueous phase, (2) organic liquids that are only sparingly soluble in water and move as a separate phase through the pore space, and (3) bacteria and viruses. The impact of these contaminants, both on the natural environment and on water supply, can be severe. Cleanup of a contaminated aquifer is a difficult and expensive task, if it is possible at all. Many basic physical, chemical, and biological issues must be resolved if society is to have the scientific basis to deal rationally with contamination in the subsurface environment. Research opportunities fall into three categories: processes, models, and field measurements.

A number of research needs are linked to general questions, discussed earlier, of the transport of reactive solutes in heterogeneous media. There is a real need for improved methods to predict solute transport over distances of 10^2 to 10^4 m, in porous and fractured rock masses, in karst terrain, and in unconsolidated sediments. Attempts to model the chemical evolution of ground water at the field scale have met with limited success. Extensive reliance is placed on experimentally determined equilibrium partition coefficients. These coefficients express, for example, the ratio of the mass of a solute species sorbed on the solid surfaces of the porous matrix, relative to the mass of the species present in the pore water. Conditions under which the use of partitioning coefficients may be adequate have not been defined clearly. There are uncertainties in transferring laboratory-determined coefficients to field settings. Little is known about the heterogeneity of the geochemical environment and the nature of its role in the transport of reactive solutes.

The subsurface behavior of synthetic organic liquids that are only sparingly soluble in water is not well known, in either the vadose or the saturated zone. Current theories describe the transport of organic liquids using concepts of multiphase flow adapted from the petroleum literature. The validity of this approach requires further examination. At the microscopic scale, the nature of interactions between the pore geometry, organic liquid, and capillary pressure-retention characteristics needs to be understood physically and described mathematically. At the mesoscale, interactions between geologic stratification and spreading patterns need to be defined, especially for organic liquids that are denser than the native ground water. Existing multiphase transport models for predicting the distribution of organic liquids in the subsurface require estimates of many parameters and functional relationships. Effective techniques to obtain these data are required.

Microbial populations indigenous to the subsurface hydrologic environment can strongly influence rates of transport and transformation of synthetic organic compounds. The nature of microbe colonies, their growth dynamics in a complex hydrologic and geochemical system, and the transformation of organic compounds into other molecular forms constitute an emerging active research area. Detailed understanding of the dynamics of this system has yet to be developed at both the microscale and the field scale. Research needs include (1) controlled experiments using multidimensional flow cells in laboratory settings, in addition to studies in field settings, (2) development of methods for parameter measurement, and (3) improvements in simulation capabilities.

HYDROLOGY AND LANDFORMS

Introduction

Geomorphic processes driven by water shape the land surfaces on which human populations and all other terrestrial biota live. In shaping the land surface, they determine the nature and distribution of hydrologic features such as river channels and soil profiles. Thus **geomorphic activity exerts a feedback on the hydrology that drives it** (Figure 3.4).

Recent technological advances give geomorphologists access to landforms below the oceans and on other planets and moons and

FIGURE 3.4 The feedback relationship between hydrology and landforms.

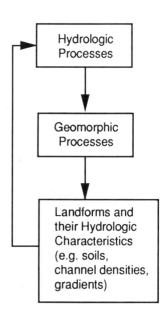

have extended the scope of this formerly earth-bound discipline. In some of these exotic environments water or ice is, or has been, an important agent shaping the planetary surface. However, most hydrologically based geomorphology concerns the surface of the earth, and thus involves processes that affect the habitability of this planet and its capacity for sustaining life in the face of sometimes radical human impacts.

During its movement over or just beneath the earth's surface, water sculpts the land and determines the nature and distribution of features such as rivers, floodplains, and soil-covered hillslopes. Societies depend on these features for their survival. Whether adequate soils survive in which to grow crops; whether rivers are navigable; whether there is magnificent scenery to inspire and delight us: these depend on the activity of geomorphic processes, driven by hydrologic processes.

Within the upper few kilometers of the earth's crust, rocks are subjected to a number of influences that weaken them, making them susceptible to erosion and transport. These influences include fracturing on a variety of scales—from microcracks to regional fracture zones—as a result of gravitation, tectonic and thermal stresses, and chemical decomposition by circulating fluids, particularly those that carry atmospheric and biogenic gases from near the earth's surface. As rock material approaches the land surface due to the stripping of overlying material by erosion, it undergoes an important sequence of physical and chemical changes and is invaded by biota. These complex changes, collectively called weathering, gradually convert the rocks to soil. The altered earth materials are the hydrologic media that determine how ground water moves, whether rainfall is absorbed by soils or flows over the ground, and how chemicals (including pollutants) are immobilized or transmitted in percolating waters. These media are also erodible and are thus involved in soil erosion, landsliding, fluvial sedimentation, and landscape evolution; they also provide substrates for human activities.

Various processes erode the decomposing materials into landforms, ranging in complexity from single hillslopes or small channels to the complex assemblages of surfaces that constitute a mountain range such as the Himalayas or an intricate alluvial plain such as the Amazon River lowland. Regional-scale landscapes develop over time scales of 10 million to 100 million years as a result of the interaction between large-scale movement of the lithospheric plates (plate tectonics) and the fluctuating climate and hydrology that drive erosion processes. Climate and hydrology in turn may be affected by the movement of continents, the growth of mountain belts, or the intensity of volcanism, as well as by changes in land use or drainage.

Within these regional landscapes, on time scales of significance to humans, geologic materials are mobilized and redistributed as poorly sorted mixtures of slightly weathered rock or as thoroughly weathered fine sediment. For example, glaciation and intense landsliding in the high Himalayas are feeding a poorly sorted mixture of sediment, ranging from boulders to silt, into steep torrents and thence to the main trunk streams flowing from or through the mountain chain. The rivers transport the various size fractions of sediment at different rates, and leave the coarser fractions behind as long-lived sedimentary accumulations (including alluvial fans, floodplains, and river terraces) in mountain valleys. The finer fractions (sand and finer particles) are transported hundreds of kilometers and deposited in adjacent basins, usually located in major downwarps of the earth's crust. Adjacent to the Himalayan chain, the vast Indo-Gangetic lowland is composed of alluvial landforms (including channel bars, floodplains, river terraces,

FIGURE 3.5 Channel network formed by ebbing tide in the Gulf of California. SOURCE: Photograph by Anne Griffiths Belt. Reprinted, by permission, from *National Geographic* 176(6), December 1989, pp. 717-718. Copyright © 1989 by the National Geographic Society.

INTERRUPTION OF MISSISSIPPI RIVER SEDIMENT FLUX

In the lower Mississippi River delta the rate of land loss has accelerated throughout this century (Figure 3.6) as delta wetlands have been inundated. The rate at which the subaerial land surface is being built up or extended into deeper water is failing to compensate for subsidence associated with the compaction of newly deposited sediment and coastal erosion. Attempts to trace the evolution of the modern Mississippi delta from historical maps and charts have concluded that if present trends continue the delta will be completely under water within 34 years.

Since the turn of the century, each of the four active subdeltas that make up most of the wetlands in the modern delta has entered a phase of deterioration. Each channel conveying sediment from the main channel to the subdelta has become shallower and more complex and therefore less efficient in distributing sediment. Meanwhile, the deltaic sediments are being compacted by dewatering, and the resulting subsidence, augmented by regional downwarping of the earth's crust and a slow rise of global sea level, eventually exceeds the diminishing rate of accretion. The problem has been aggravated because the rate of sediment supply to the delta has decreased by approximately 50 percent, primarily as a result of trapping of sediment in large reservoirs in the Missouri and Arkansas river system and the diversion of 50 million tons per year out of the Mississippi channel into the Atchafalaya distributary channel (Figure 3.7).

Understanding the interaction of these processes is a necessary basis for any plan to adjust to and ameliorate the loss of wetlands in the Mississippi or other deltas.

---

alluvial fans, river channels, and lake basins) that are being formed and destroyed as sediment from the mountains is deposited and later eroded again and moved downstream.

As a land surface is eroded, water is concentrated along depressions either by surface runoff from higher parts of the landscape or by ground water discharging into the depressions. The result is a stream network in which water flows faster and deeper than over the surrounding landscape and has a greater capacity for transporting sediment evacuated from hillslopes. The resulting branched, hierarchical network of channels exhibits a high degree of regularity and spatial organization (an example is shown in Figure 3.5). The land surface draining to any point on a channel network is defined as the drainage basin of the stream at that point. The geometry and sedimentology of channels are scaled by the magnitude of the water flows that they receive from their drainage basins and also by the magnitude

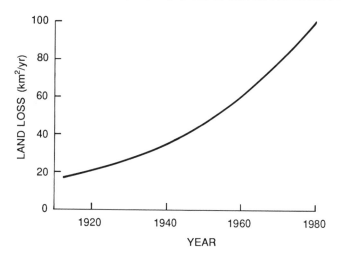

FIGURE 3.6 Land loss in the lower Mississippi River delta of Louisiana. SOURCE: Reprinted, by permission, from Gagliano et al. (1981). Copyright © 1981 by S. M. Gagliano.

and grain size of the sediment load that the water transports from upstream. Since drainage area and thus water and sediment fluxes change in a regular manner along the stream network, so also do the characteristics of channels. **Perturbations of the water or sediment fluxes from the drainage basin, as a result of environmental change or human activity, cause changes in channel and valley floor characteristics, often with widespread effects on human habitat.**

Some Frontier Topics

Prediction of Landscape Evolution

Most geomorphological research has been aimed at understanding the nature and chronology of landscape development, and it has yielded many important insights concerning the large-scale form and dominant erosion processes of most of the earth's regional landscapes. However, almost all of this work has been qualitative or empirically quantitative, and explanations generally are given a posteriori. At the other end of the spatial scale, the past 40 years have witnessed improvements in understanding individual erosion processes and predicting their effects on the sediment yield and shape of individual hillslopes, including the effects of varying climate and land use. There is still much to be learned about the mechanics of individual

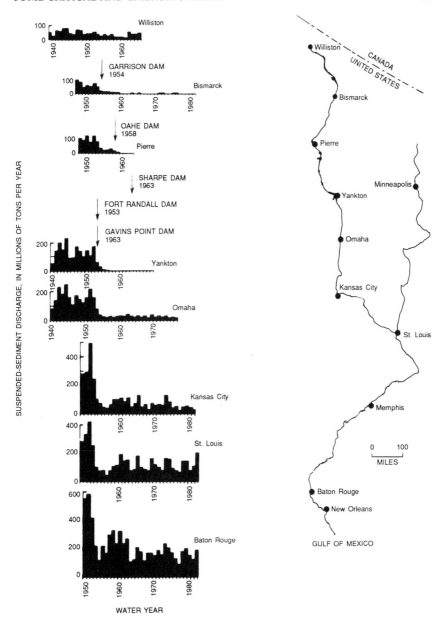

FIGURE 3.7 Annual discharge of suspended sediment at six stations on the Missouri River and two stations on the Mississippi River showing the effects of reservoirs on downstream sediment loads, 1939 to 1982. SOURCE: Reprinted from Meade and Parker (1985) courtesy of the U.S. Geological Survey.

processes of erosion and sediment transport, including such wide-spread and important processes as the interaction of raindrop impact and surface runoff in causing soil erosion, the buildup of stress as a root-reinforced soil approaches the threshold of landsliding, and the initiation and motion of catastrophic landslides and debris flows. These hillslope sediment transport mechanisms now are being studied with new instruments for measuring turbulence, strain, and other characteristics of deforming media, and with faster computers to analyze geometrically complex situations. Understanding these processes is important for forecasts of natural hazards and sediment yields in the context of anticipated changes in climate and land use, or in catastrophic geologic events such as volcanism.

How are water, sediment, and nutrients exchanged between river channels and their floodplains?

Both the understanding of mechanics and the capacity for prediction of geomorphic processes decrease as one moves to interactions between processes on larger scales, but the problems and prospects are exciting. For example, research on the exchanges of water, sediments, and nutrients between river channels and their floodplains is in its infancy, despite long-standing assumptions that the fertility of floodplains is the result of nutrient-rich sediment from certain favorable rocks. Such issues are brought sharply into focus by major engineering projects (such as the High Aswan Dam) that interrupt the sediment flux and diminish overbank flooding, but the fundamental research necessary for quantitative prediction of the effects of such projects on floodplain sedimentation is just beginning.

How do regional-scale landscapes react to changes in climate and the biosphere?

Interactions between erosion and sedimentation processes are even more complex and important when one considers how regional-scale landscapes react to changes in climate and the biosphere, and therefore to anticipated global warming and land use changes. A landscape is an ensemble of hillslopes with different shapes that shed sediment episodically into river channels bordered by floodplains where the sediment is transported and stored for various lengths of time. The sediment includes plant nutrients and organic and toxic substances, which can be altered and separated during storage and transport. With the insights provided by process studies and with the aid of increasing computing power, geomorphologists are developing useful simplifications of erosion-sedimentation models to make regional-scale prediction computationally tractable while retaining a useful degree of realism that could be checked against field observations or geologic records.

Prediction of landscape-scale erosion and sedimentation requires viewing landforms in a wider context, including soil and vegetation covers. Changes in the condition of land surfaces therefore depend on the link between erosion and the regional soil-water balance, vegetation cover, and land use. Impetus for such modeling of large-scale processes will probably soon emerge from the capacity to link general circulation models, regional-scale water balances, and erosion processes.

Because of the need to predict changes over large regions, one must consider the response of entire river and valley floor networks to sediment influx and therefore the link between hillslope and floodplain sediment. Current methods do not take account of the storage of sediment in floodplains and alluvial fans, which modulates sediment-yield response most strongly during periods of environmental changes. An important task is to develop generalizations about storage and remobilization from the field studies now being undertaken with the aid of radiometric dating and chemical tracers. In the process of accumulating this understanding, much will be learned about

- the basis for the fertility of floodplains,
- the geomorphologic and hydrologic basis for the ecological diversity of floodplain flora, and
- the fate of toxic chemicals sequestered in sediments and concentrated at drainage basin outlets.

Development of Drainage Basins and Channel Networks

How should one model the mutual dependence of channel and hillslope surfaces as a key factor in theories of drainage basin evolution and channel network formation?

River channel networks—like leaf veins, blood vessels, and trees—are branching networks that transport fluids to or from three-dimensional surfaces. Biological and physical scientists have long been fascinated

98

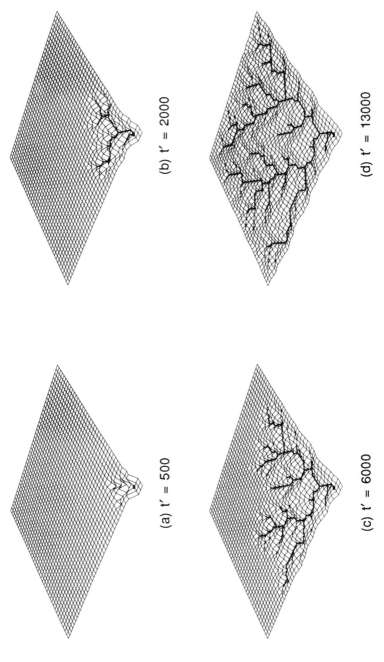

(a) t′ = 500

(b) t′ = 2000

(c) t′ = 6000

(d) t′ = 13000

FIGURE 3.8 Simulation of drainage basin development with time (t′ = dimensionless time). SOURCE: Reprinted, by permission, from Willgoose et al. (1989). Copyright © 1989 by Ralph M. Parsons Laboratory, Massachusetts Institute of Technology.

by questions of form and function, such as what causes leaf cells to differentiate into stems and matrix and how topography differentiates into channels and hillslopes. Some geomorphologists have studied the mechanics of the erosion processes that form channels in particular hydrologic environments; others have attempted to define more generally the necessary and sufficient conditions for channel incision on an erodible surface; and still others have formulated a mathematical description of channel networks as outcomes of a random selection process. Recent studies have focused on the differentiation and mutual dependence of the channel network and the hillslopes that constitute most of the drainage basin surface.

Researchers studying a variety of tree-like structures have concluded that they all have some common growth characteristics. In hydrologic terms these characteristics require the existence of an erosion regime that depends strongly on the magnitude and frequency of water flow per unit width, leading to a progressive change in the processes and forms of sediment transport from hillslopes to rivers. The position of channel heads is determined by the long-term balance of sediment movement. Upstream, valleys tend to fill with sediment, whereas beyond a threshold distance from the drainage divide channel incision becomes established. The location of channel heads is controlled through an interaction between gradient and catchment area, which together influence surface and subsurface hydrology and sediment transport. Competition for runoff, and therefore drainage area, between growing channel networks leads to a land surface that is differentiated into hillslopes and channels connected in a regular, dendritic, space-filling pattern. The change of the sediment transport mechanism from hillslope to channel may be modeled through the attainment of a threshold value by the function describing the mechanics of the particular erosion process (such as surface erosion, seepage erosion, or mass failure) that extends channels in the particular landscapes to which the model is applied. Pioneering work in the quantitative description of basin evolution was carried out along those lines by Smith and Bretherton (1972).

Three-dimensional landscape evolution models incorporating ideas of this form first require an equation describing the sediment balance at any point. The sediment transport in this balance is nonlinearly dependent on the local water discharge and slope. The resulting equations are solved numerically in space and time. To initiate the process, a small random elevation perturbation is assigned to every point in the catchment. Figure 3.8 shows examples of networks generated in this way. Analysis of planar and elevation properties of many of these simulations indicates behavior consistent with that observed in nature. The generated networks are extremely sensitive to initial conditions. This is reminiscent of chaotic behaviors that may look like the outcome of a random process, although in fact they may arise from completely deterministic equations (see the last section of this chapter, "Hydrology and Applied Mathematics").

Important questions remain to be addressed, such as what the parameters are that control the growth and differentiation of the channel network and how they depend on climate and topography.

Three-Dimensional Network Geometry and Hydrologic Basin Response

---

What hidden unifying principles lie in the three-dimensional geometry of channel networks and how are they related to the hydrologic response of a basin?

A key idea introduced by Horton (1945) and modified by many others is that of the ordering of a channel network. In the 1960s, Shreve (1966) provided an analytical framework under which Horton's laws of network regularity could be studied. In spite of many efforts, the analytical theory of channel networks has remained almost purely planimetric. A challenging and crucial problem is the incorporation of relative elevations and their ties with the planar characteristics of the drainage network.

In the 1970s, mainly through the work of Rodriguez-Iturbe and Valdes (1979) in Venezuela and Kirkby (1976) in England, it was shown that the topology and geometry of the channel network are intimately linked with the hydrologic response of drainage basins and particularly with the flood hydrograph that results from storms occurring over the basin. The geomorphologic unit hydrograph (GUH) derived from this work is a first step toward rigorously connecting the unit impulse response function of a basin with its geomorphological characteristics. Nevertheless, the connection is only related to the planar structure of the basin; the altitude dimension is still missing in the efforts to link the structure of the channel networks with the hydrologic response. Moreover, the GUH and other similar schemes are only routing procedures, since they work with effective rainfall and do not address the question of runoff generation. Because of the many circuits of reciprocal control between the system of hillslopes and the drainage network, one could expect that the understanding of the three-dimensional organization of the network will also hold the key for the organization of the runoff production process in natural basins.

Stream networks are the result of dissipation of energy in the environment by water flowing downhill. This simple observation shows that drainage networks, beyond their purely topological and planimetric attributes, are dissipative physical systems. The ability of drainage networks to erode into landscape is affected by a balance between the ability of the network to do work versus the resistance of the landscape to erosion. Therefore it is reasonable to suspect that the three-dimensional geometrical patterns manifested by drainage systems record a variety of states and conditions, including the total flux of material through the system (related to climate), the distribution in time of such fluxes (related to weather), and the rates of change of fluxes.

Recent research shows that a promising avenue in understanding the analytical structure of the three-dimensional geometry of river networks is to identify the precise nature of the scaling property in the network geometry. As explained in the last section of this chapter, scaling refers to an invariance property of the probability distributions of physical variables under a change of spatial scale. Although predictions based on the simple scaling hypothesis are quite good, the data also show a need to generalize to multiscaling. Another important issue in this context is the determination of the scaling exponents, appearing as parameters in a scaling theory of river network geometry, from physical considerations about climate, geology, landforms, and so on. Moreover, a physically based determination of these scaling exponents is expected to identify some general physical principles, e.g., uniform distribution of energy expenditures in channel networks, as a basis for discovering the signature of river basin dynamics in the geometry of channel networks.

Investigations outlined above involve only the spatial variability and neglect the issue of time. In this sense they constitute only the first step toward developing a mathematical theory of river basins. Extensions of such spatial theories to space-time evolutions, and the solution of the novel and important problems that would arise through such generalizations, remain among the major long-term research goals.

Interpreting Records of Environmental Change

What new techniques can be used to retrieve the information on potential geomorphic alterations and climate changes that is stored in sediments?

Geological and historical records of the postglacial era indicate that widespread changes in erosion and sedimentation resulted from subtle climatic changes during the past 10,000 years. These climatic fluctuations alter plant cover, frequency distributions of rainstorms, the presence and rates of snowmelt, evapotranspiration, and other factors that in turn control the condition of land surfaces (e.g., their erodibility, moisture status, and radiative properties) and the response of rivers and valley floors to changes in runoff and sediment supply.

For example, vast quantities of sediment were deposited tens of meters thick over thousands of kilometers of valley floor in subhumid regions throughout the world as a result of climatic changes during postglacial times. These valley deposits have been trenched and re-plenished repeatedly by streams. In some areas the sequence of erosional and depositional events has been complicated by the effects of land use. One of the most studied cases is in the southwestern United States, where there is a controversy about the relative roles of climatic change and land use in initiating a regional cycle of arroyo entrenchment late in the nineteenth century. Throughout the world, there are many other such records of the morphological and sedimentation effects of subtle changes in climate and/or land use on stream channels, valley floors, and the soil cover of hillslopes. In some localities, chemical tracers and isotope distributions also provide information about sources of sediment and rates of accumulation.

If anthropogenic disturbances cause future climatic changes with rates and magnitudes at least equal to those in postglacial times, it is reasonable to anticipate that there will again be widespread alterations of the vegetal and hydrologic condition of subhumid land surfaces, and changes in the magnitude of hillslope erosion relative to the sediment transport capacity of trunk streams. If the sediment yield of the hillslopes exceeds the transport capacity of the streams, sediment will accumulate and raise valley floors. Such changes will affect riparian ecosystems, agricultural land, irrigation systems, and urban areas. Some means of anticipating these effects is needed, and the alluvial deposits of valley floors provide the most useful indicators of analogous changes in the past. However, it is difficult to make predictions of the effects of future environmental changes on erosion and sedimentation, and their enormous influence on human welfare, from qualitative interpretation of sedimentary records. A major challenge for hydrologists concerned with sediment redistribution is the quantitative interpretation of morphological and sedimentological records of environmental changes, and the use of these records for testing prediction models or simply for making more secure predictions by analogy with past changes.

GROVE KARL GILBERT
(1843-1918)

G. K. Gilbert was one of the earliest earth scientists to combine a theoretical background in the basic sciences with extensive, thorough field observation to develop general theories of landform evolution, including the roles of hydraulics, material properties, and solid-earth geophysics. Although his field-based theories were expressed qualitatively, they were rigorously framed and testable, and when rediscovered in the 1940s, they stimulated many quantitative field and laboratory geomorphological studies.

Gilbert was the first scientist to emphasize the relationship between landscape forms and the mechanics of their generative processes, and he emphasized the interrelationships of river channels, hillslopes, and nested sets of drainage basins in controlling how sediment is moved and the earth's surface transformed. His preference for analyzing the near-equilibrium operation of landforms such as river and hillslope profiles and beaches presaged modern work on simultaneous adjustment and negative feedback in landform evolution.

Throughout his career, Gilbert's attempts to develop theory led him to focus gradually on studies in which the influence of particular variables could be isolated. Early in his career, this was achieved through judicious selection of field sites—sites Gilbert could analyze in a rigorous, theoretical manner. Later, he was asked to study an inadvertent geomorphological "experiment" in which vast quantities of sediment were sluiced into the Sacramento River system and San Francisco Bay as a result of hydraulic mining. Gilbert therefore had access to resources for detailed topographic surveys, granulometry, and discharge data. In conjunction with these studies he also initiated the first set of laboratory flume experiments on sediment transport conducted by an earth scientist concerned with the operation of river channels.

In addition to pioneering geophysical approaches to geomorphological theory, experimental geomorphology, and the application of that science to sedimentation problems, Gilbert conducted studies of lunar origin and lunar landforms, and generally operated in a manner that, a century later, has proven remarkably prescient, powerful, and immediately interesting.

Examples of other regions where such predictions could be useful include the following:

- the Amazon basin, where the effects of land use changes on river sedimentation have not yet been well documented or predicted;
- the Mississippi basin, where impoundment of sediment in reservoirs has reduced the sediment supply to the lower valley and delta

and caused widespread erosion, but where important climatic changes
may occur that could affect both river flow and sediment; and

• the Ganges-Brahmaputra basin, where there are major uncertainties about the future roles of Himalayan erosion, land use, in-
channel sedimentation, intensified monsoonal weather patterns, and
rising sea level on the geomorphology and therefore habitability of
the alluvial zone of Bangladesh.

The keys to making and testing such predictions lie in the physical
and chemical characteristics of alluvial sediments. The thorough investigation of these sediments, combined with process-based mathematical modeling of their accumulation, constitutes the best source
of information for anticipating changes in sediment redistribution.
Researchers in this field need training and equipment for using isotopes
and other chemical tracers, as well as skills in stratigraphy, sedimentology,
and geomorphological modeling.

HYDROLOGY AND CLIMATIC PROCESSES

Introduction

Water has physical properties that significantly influence the global climate. Large amounts of heat are associated with the phase
transitions between water vapor, liquid, and ice. The release of latent
heat during surface evaporation is the major mechanism through which
much of the absorbed solar energy at the surface is transferred to the
atmosphere. The evaporated water then moves within the atmosphere
in the form of water vapor and clouds, eventually condensing into
precipitation. The released latent heat drives air motions important
to the atmospheric general circulation.

Water vapor is also the most important of the greenhouse gases,
which efficiently absorb and then return the thermal radiation emitted
by the heated earth surface. In this manner, vapor in the atmosphere
regulates the temperature regime at the surface as well as in the
atmosphere.

The heat capacity of water makes the oceans an important regulator of global climate. At high latitudes, sea ice and snow exert significant influences on global climate because of their enhanced capability to reflect incident solar radiation.

Given the myriad ways in which the occurrence and abundance of
water can influence the climates of the world, some **major goals for
hydroclimatologic research must include (1) improved understanding of the interaction between the hydrologic cycle and the general
circulation of the coupled ocean-atmosphere system, and (2) eluci-**

dation of the role of this interaction in maintaining climate and influencing its variabilities.

An important first step toward these goals is the quantitative determination of water fluxes at the ocean and land surfaces as well as in the atmosphere. To address this need, a plan is being developed for comprehensive in situ and remote observations of the land surface and the atmosphere. This plan, put forward by the World Climate Research Program (WCRP) of the World Meteorological Organization and the International Council of Scientific Unions, is known as the Global Energy and Water Cycle Experiment (GEWEX), and it is described in Chapter 4.

Remote sensing methods, also described in Chapter 4, have already been developed to assess such variables as surface soil moisture, surface temperature and reflectivity, vegetation cover, large-scale patterns of rainfall and cloudiness, and vertical profiles of water vapor in the atmosphere. We still need to improve and verify calibration techniques, however.

A major tool for the study of climate and the hydrologic cycle is the numerical atmospheric general circulation model (GCM). During the postwar period, GCMs evolved from simple dynamical tools to help forecast daily wind patterns into more complicated mathematical algorithms that capture the dynamics and linkages between general atmospheric circulation and the hydrologic cycle. Results of computer simulations with GCMs can, for example, illustrate the role of land surface fluxes from a local area on the regional climate (Figure 3.9). Numerical models can serve as laboratories for evaluating the impact of such phenomena as deforestation in the Amazon and extensive droughts in Africa.

A comprehensive approach involving both observation and modeling of the coupled ocean-atmosphere-land surface system is essential for the effective study of climate.

Some Frontier Topics

Diagnostic Study of the Global Water Balance

How can general circulation models simulate the quantitative structure of the global hydrologic cycle and help to evaluate its effect on climate?

The hydrologic cycle is imbedded in the general circulation of the atmosphere. The capability to simulate and predict the general circulation of the atmospheric fluid is thus of crucial relevance to hydrology.

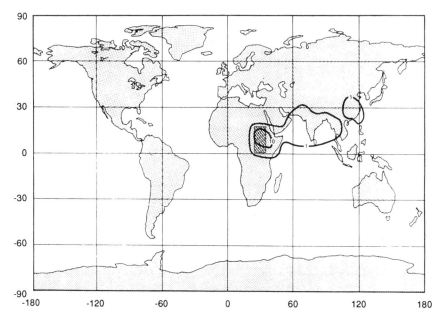

FIGURE 3.9 Contours showing the general circulation model estimate of precipitation (mm) originating from evaporation on the Sudd region (shaded area) of Sudan for 30 days in July. SOURCE: Reprinted, by permission, from Koster et al. (1988). Copyright © 1988 by Ralph M. Parsons Laboratory, Massachusetts Institute of Technology.

General circulation models have an impressive history in atmospheric science not only as operational tools in weather forecasting, but also in climate research; only recently, however, have GCM modelers and hydrologists discovered that they have interests in common.

To the GCM modeler, storage and flow paths of water in the ocean-atmosphere-land and surface-ice system are central to the proper simulation of weather patterns and climate. Coupled land surface water and heat balances have strong influences on the distributions of water vapor, temperature, and flow in the atmosphere. Other significant hydrologic factors in the model include the extent of snow and ice cover, which affects the surface heat budget. The accurate representation (or parameterization) of the processes involved in the hydrologic cycle is essential for successful general circulation modeling.

To the hydrologists, GCMs offer a fresh approach to two main principles of hydrologic science—the hydrologic cycle and the water balance. The GCM can serve as the experimental apparatus for studying the spatial and temporal patterns in the atmospheric and surface branches of the hydrologic cycle. Furthermore, with GCMs the large-scale atmospheric water balance and surface hydrology may be analyzed for their influences on regional climates; the role of land surface-atmosphere feedbacks in maintaining climate and controlling its sensitivity may also be examined.

One of the challenging research tasks facing climate modelers is the elucidation of the processes that take place at the surface of major continents and influence the distributions of climate and hydrologic conditions. For example, one can ask what role land surface-atmosphere interaction plays in maintaining the arid and semiarid regions of the world. We know that the radiation energy reaching the continental surface is ventilated through evaporation and the upward flux of sensible heat into the atmosphere. When a continental surface is dry, the ventilation through sensible heat becomes larger than that of evaporation, thereby raising temperature and decreasing the relative humidity, cloudiness, and precipitation in the lower troposphere. The reduction in cloudiness, in turn, increases the radiation energy reaching the continental surface, and thus increases potential evaporation. Both the reduction of precipitation and, to a lesser extent, the increase of potential evaporation further deplete soil moisture from the dry continental surface. This description indicates that the land surface-atmosphere interaction has a positive feedback effect on the aridity of a continent.

The role of the land surface-atmosphere interaction is critical in maintaining climate and controlling its temporal variation. Thus a realistic modeling of the land surface process is essential for the successful simulation of climate by a GCM and for its use in regional hydrologic studies. However, it is extremely difficult to determine realistic rates of evaporation and runoff over each grid box of a GCM, given that the dimensions can vary from a few kilometers to several hundred kilometers and that each grid box contains highly heterogeneous surfaces. For example, the evaporation rate is influenced by vegetation in each grid box through physiological processes involving stomata and a root system. Such an effect is not explicitly incorporated into most GCMs. Devising reliable parameterizations of evaporation and runoff from a macroscale grid box with a heterogeneous surface is one of the most challenging tasks for reliable simulation of climate and the hydrologic cycle. Given that GCM grids are typically 10^4 to 10^5 km^2, the significant effects of spatial heterogeneities on surface hydrology need to be represented. Similarly, procedures need to be developed to link regional heat and water balance to those at sparse GCM grid points.

To develop and validate a more realistic land surface model for a GCM, a series of field experiments was planned by the WCRP. Called

ATMOSPHERIC GENERAL CIRCULATION MODELS

Early in the twentieth century, L. F. Richardson proposed that, since the atmosphere is a moving fluid, its motions could be predicted using the basic equations of hydrodynamics. Unfortunately, the sheer size of the problem made the number of necessary calculations too large to be carried out on desk calculators of the time. After World War II, the development of electronic computers vastly increased the calculating power that could be directed to solving the equations of atmospheric motion. Stimulated by the success in numerical weather forecasting, N. A. Phillips added the effect of thermal forcing to the hydrodynamical equations and pioneered using such general circulation models (GCMs) for the study of the atmospheric circulation. In the late 1960s the hydrologic processes were incorporated into GCMs, resulting in successful simulation of the basic behavior of the hydrologic cycle and its effect on climate.

A general circulation model predicts the changes of the atmospheric flow field and temperature based on the equations of hydrodynamics and thermodynamics, respectively. Based on the equation of radiative transfer, it takes into consideration the heating and cooling caused by solar and terrestrial radiation. The three-dimensional distribution of atmospheric water vapor is predicted by considering the effects of advection, condensation, and moist convection. The temporal variation of soil moisture usually is determined from a simple water budget with contributions from rainfall, evaporation, snowmelt, and runoff. The temperature of the continental surface is determined from heat balance calculations. Figure 3.10 demonstrates the GCM's capability to reproduce the observed values of temperature, precipitation, and evaporation averaged over latitude circles. Unfortunately, however, at the present state of GCM development, reproduction of the observed regional variability of these averaged quantities is at best only qualitatively accurate. This is illustrated for runoff by Figure 3.11. This deficiency of GCMs must be removed before they can become practical tools for national and regional decision making.

By the use of GCMs, some success has been achieved in identifying various factors that play a major role in maintaining climate. One such factor is major mountain ranges, such as the Rocky Mountains and the Tibetan Plateau. They act not only as obstacles for air flow, but also as elevated heat sources, thereby exerting profound influences on general circulation and the hydrologic cycle in the atmosphere. By comparing simulations with and without these major mountain ranges, the roles of the Tibetan Plateau and the Rocky Mountain ranges in shaping the jet stream, developing the rainy Indian summer monsoon, and maintaining the extreme, dry conditions over the Gobi Desert have been investigated.

General circulation models also have been used to explore the role of oceans in influencing the geographical distributions of hydrology and climate at the continental surface. Since oceans can be thought of as the ultimate sources of water, their importance in maintaining the hydrologic cycle and land surface hydrology cannot be overemphasized. In addition, the heat transport by ocean currents affects the sea surface temperature, thereby influencing the evaporative supply of water into the atmosphere. For instance, we know that the thermal inertia of oceans is responsible for the seasonal reversal in the land-sea contrast of surface temperature, thereby inducing the major monsoons.

Because of their large thermal inertia, oceans also play an important role in long-term climate change. For example, the rate of future climate change caused by greenhouse gases will be determined in part by the penetration of heat energy into the deeper layers of oceans. For the study of long-term climate change, it is essential to use a climate model in which an atmospheric GCM is coupled with an ocean circulation model.

the Hydrologic-Atmospheric Pilot Experiments (HAPEX), this series measured basic components of water and heat budgets and other relevant quantities over a land area of approximately $(50 \text{ km})^2$ to $(100 \text{ km})^2$. HAPEX is described in detail in Chapter 4.

For the validation of a climate model, it is also necessary to determine more accurately the thermal and water balance of the global atmosphere and the exchange of energy and moisture between the global atmosphere and the land, ocean, and ice-covered surface based on the comprehensive observations from satellite and ground stations. One of the main objectives of GEWEX is to accomplish this by a comprehensive strategy combining both observational and modeling approaches (see Chapter 4).

What are the states and the space-time variabilities of the global water reservoirs and their associated water fluxes?

The total amount of water in the earth-atmosphere system must remain relatively constant; therefore, an imbalance in the three exchange processes of precipitation, runoff, and evaporation must be offset by changes in the storage reservoirs of the ocean, land, cryosphere, and atmosphere. There is constant exchange between the reservoirs via these processes, but the time scale of the exchanges varies greatly among the many components of the system. An understanding of

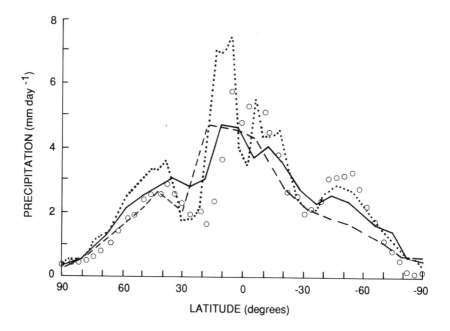

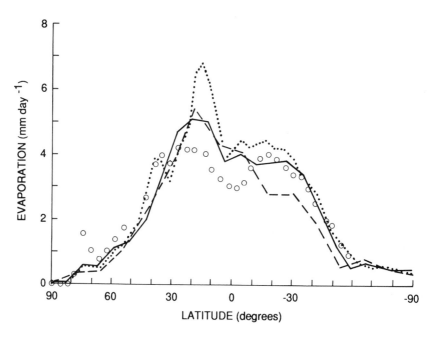

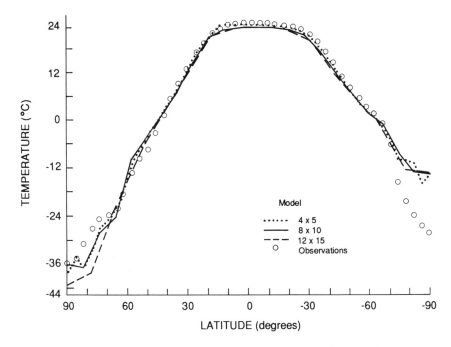

FIGURE 3.10 Results from the general circulation model of NASA's Goddard Institute for Space Studies. The model, with nine vertical layers and various horizontal resolutions, has been integrated for 5 years, and the seasonal zonal values for temperature, precipitation, and evaporation are averaged. SOURCE: Reprinted, by permission, from Hansen et al. (1983). Copyright © 1983 by the American Meteorological Society.

the global water balance and its variability requires accurate assessment of both reservoirs and fluxes.

The emergence of an accurate global picture of the water balance has been hindered by several factors:

1. Observations are spotty in their spatial coverage, being concentrated where the human and financial resources have been available. Underdeveloped areas such as arctic and alpine regions, deserts, the tropics, and, of course, the oceans, have been neglected. Attention must be given to filling these gaps by remote sensing and unattended devices.

2. Observations are of uneven quality owing to the differing local and national standards. The World Climate Data Program (WCDP) is addressing this issue.

3. Observations are limited in character primarily to the quantities of traditional engineering interest, namely, precipitation, streamflow,

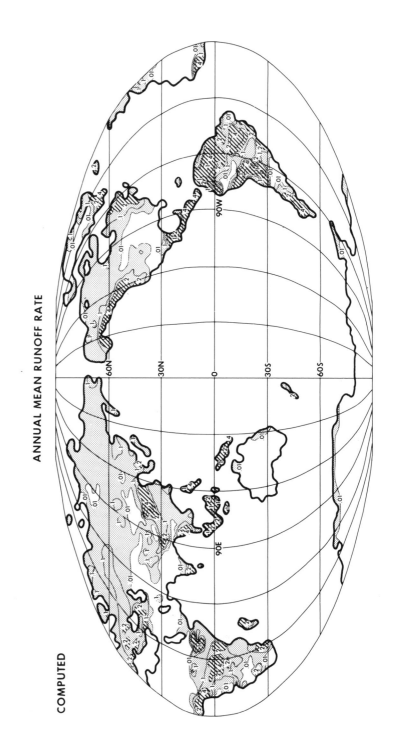

ANNUAL MEAN RUNOFF RATE

COMPUTED

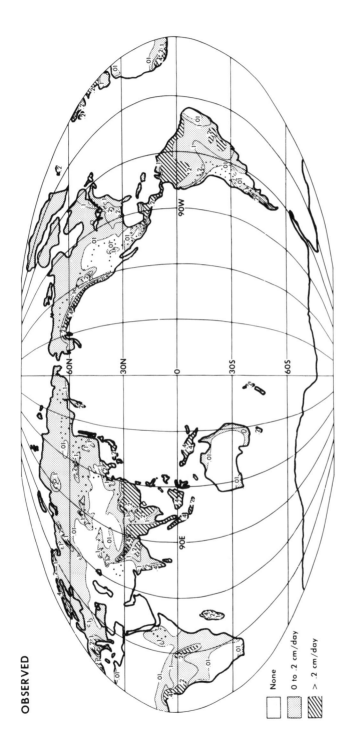

FIGURE 3.11 Global distribution of annual mean rate of runoff (centimeter per day) simulated by the global model of Manabe and Holloway (1975) (top) compared with an observed distribution (bottom) based on data derived from L'vovich and Ovtchinnikov (1964). Small-scale features of contours have been removed subjectively. SOURCE: Top—Reprinted, by permission, from Manabe and Holloway (1975); copyright © 1975 by the American Geophysical Union. Bottom—Reprinted, by permission, from L'vovich and Ovtchinnikov (1964); copyright © 1964 by the American Geophysical Union.

OBSERVED

None

0 to .2 cm/day

> .2 cm/day

FROM DESERT TO RAIN FOREST

Over recent geologic time, highly visible changes in the global water balance have occurred—the rise and fall of sea level, the advance and retreat of continental ice sheets, and the formation and desiccation of major lakes and waterways. At the peak of the last Ice Age, some 18,000 years ago, when as far south as Kentucky some 1,500 m of ice covered the land, increased aridity generally prevailed in lower latitudes. The Sahara Desert advanced nearly 1,000 km southward toward the present-day rain forests, virtually burying the river systems of West Africa. The rain forests of Africa, South America, and probably Asia all but disappeared, retreating to a few anomalously humid highland regions. At the same time, vast lakes covered much of the western United States, the remnants of which are seen in the salt flats of California, Nevada, and Utah. Millions of tons of glacier ice on land tied up a 100-m layer of the world oceans, exposing huge areas of the continental shelves. Within a few thousand years, the glaciers melted, sea level rose, and savannas and lakes replaced most of the Ice Age deserts. About 5,000 years ago, the deserts, including the Sahara, had nearly vanished. In Africa and Australia, lakes as large as Lake Chad expanded to many times their present size; some Rift Valley lakes were up to 150 m deeper than they are now. Neolithic fishhooks found in the central Sahara bear witness to the existence of Saharan lakes suggested by geological deposits.

In the past, evidence of such long-term changes derived principally from geologic studies and occasionally palynological and archaeological information. Advanced satellite technologies now are providing innovative ways to study past conditions on a large scale. One example is a surprising discovery made from the microwave radar system aboard the first shuttle flight. In hyperarid environments, microwaves penetrate several meters below the surface before being reflected spaceward. The shuttle radar showed subsurface patterns in the western desert of Egypt that proved to be a long-buried waterway, along the banks of which a number of archaeological sites were unearthed. The channels were established in Tertiary times, millions of years ago, but flow was periodically reestablished during the Ice Age, and the region reoccupied by early humans. Speculation that these now-dry channels once linked the Niger and Nile rivers, as was the tradition in African legends, was reinforced with the discovery of a fossil fish skeleton in the driest core of the desert. This fish, the "el Capitán," now inhabits the Niger and Nile, but few other, rivers; its presence in the Sahara strongly suggests a past link between these two river systems.

Although such knowledge of the past helps us to understand the course of human history, it may also prove key to solving many current problems. A large portion of our ground water resource, which today is being depleted and contaminated, is a remnant of these past humid periods. Facing the potential for global climate change, we must know the likely range of expected conditions: rainfall, flow of rivers, size of lakes, and sea level changes. It also is important to know whether, on a global scale, the partitioning of water among the various reservoirs and the rates of transfer between them (i.e., the fluxes) will change significantly. In other words, is the total global precipitation constant and merely geographically redistributed over time? Or does the efficiency of the hydrologic cycle, i.e., the global rates of precipitation and evaporation, change through time? Is the amount of atmospheric water vapor and clouds constant or variable? Studies of long-term changes of the global water balance help to provide answers to such questions, which are posed by the possibilities of human-induced climate change through greenhouse warming or other factors.

and surface water reservoirs. Observations of the atmosphere (particularly the troposphere), soil moisture, ground water, and evaporative fluxes have been relatively neglected. Advances in technology are badly needed in these cases.

4. Observations are largely unavailable in the form of coordinated, homogeneous, global data sets. Again, the WCDP is attacking this problem.

In addition, most measurements of parameters such as rainfall, soil moisture, and evapotranspiration are point measurements from which we attempt to extrapolate large-scale fields. Unfortunately, these parameters are highly variable in space over small distances, and assessing them on a regional or global scale has proven difficult.

A major component of climatic change involves the interactive relationship between global climate and regional-scale hydrology. On the regional scale we lack knowledge of such basic questions as the geographical source regions of atmospheric moisture, the net moisture transport into or out of the region, and whether or not local evaporation is a significant source of atmospheric moisture supply. Fortunately, as has been discussed above, computer simulations are now helping to answer these questions (e.g., see Figure 3.9). Such knowledge is important to validating numerical models before they can be used to evaluate the impacts of such phenomena as deforestation in the Amazon and the extensive droughts in Africa.

In summary, scientific challenges pertinent to global water balance include accurate assessment of the atmospheric reservoir and its exchange processes, determination and prediction of long-term changes

in the global water balance, and assessment of the human impact on the water cycle.

To fully understand the global water balance, its regional interactions, and long-term fluctuations, numerous aspects of the atmospheric water cycle require further study. These include (1) the global distribution and seasonal variation of rainfall, evaporation, and runoff, (2) the temporal variations of rainfall regionally and for the earth as a whole, (3) moisture transport pathways and storage in the atmosphere, (4) global space-time variations of cloudiness, (5) the contribution of the land surface, especially tropical forests, to atmospheric moisture supply, and (6) the residence time of moisture in the ground and atmosphere. A scientific frontier in the investigation of the atmospheric reservoirs and fluxes is the development and refinement of appropriate satellite measurement techniques. In particular, techniques for estimating rainfall and evapotranspiration, assessing wind fields and moisture profiles, and monitoring clouds would help to develop a detailed picture of the atmospheric component of the global water cycle and its role in global climate. This is a prerequisite to deriving estimates of long-term changes in the global water balance, an area of ever-increasing concern in view of the prospect of human-induced worldwide changes in climate.

Surface-Atmosphere Interaction

What are the roles of atmospheric dynamics and surface processes in determining the variabilities in climate?

The factors determining the variability of climate can be divided into two types: factors that affect the internal dynamics of the atmosphere, and surface or boundary factors external to the atmospheric system. The boundary forcing is exerted as the fluxes of heat, water vapor, and momentum between the surface and the atmosphere. It is influenced by the fluctuations of sea surface temperature, soil moisture, snow and ice surfaces, and vegetation. The atmosphere is relatively transparent to solar radiation. Therefore its primary direct source of energy is the earth's surface, which absorbs solar radiation and transforms it into forms of energy that are readily transferable to the atmosphere. Thus boundary forcing from the surface is an important determinant of climate.

In contrast to rapidly fluctuating internal atmospheric conditions,

surface characteristics (e.g., sea surface temperature and soil moisture) vary much more slowly. This suggests a potential mechanism
for predicting atmospheric behavior on long time scales. Thus, with
adequate understanding of the surface forcing and accurate characterization of it in predictive models, seasonal or multiseasonal forecasts might be realized.

The ocean's influence on the interannual variability of weather
and climate has been recognized for some time. The El Niño/South-
ern Oscillation phenomenon, which at irregular intervals evokes a
global pattern of anomalous weather conditions, is a prime example.
An opposition or seesaw of sea surface temperatures in the eastern
and western Pacific induces major temporal fluctuations of precipitation,
soil wetness, and river runoff over Australia, Central and South America,
and other parts of the world, and these can persist for a year or more.
Ocean-atmosphere interaction is fundamental to this oscillation. **Although the mechanisms of this interaction lie outside the committee's
definition of hydrologic science, the results of the interaction are
crucial to the hydrologic cycle at global and regional scales.**

What is the interactive relationship among rainfall, cloudiness,
soil moisture, surface temperature, reflectivity, and vegetation
cover?

Many processes by which the land and atmosphere interact involve the exchange of energy (e.g., sensible heat, latent heat, and
radiation), mass (e.g., moisture and aerosols), and momentum.
Modification of surface temperature and heat balance by changes of
albedo, soil moisture, and vegetation cover most directly influence
the energy exchange. The surface also transfers mass to the atmosphere
in the form of water vapor and dust. This indirectly redistributes
energy in the form of latent heat and alters the patterns of atmospheric heating. Changes in albedo, soil moisture, vegetation, and
evapotranspiration play key roles. **Land surface processes have important effects not only on the availability of moisture, but also on
patterns of surface and atmospheric heating, which in turn influence
the dynamic behavior of the atmosphere.**

If altering the land surface can influence the hydrologic cycle, we
must know in what locations, under what conditions, and on what
time and space scales. Likewise, we must establish which of the
potential processes are most important and on which time and space
scales. Only then can we determine if feedback between the land

surface and atmosphere is sufficiently large that drought or wet conditions might be self-reinforcing, leading to multiyear persistence of the moisture state.

Charney (1975) suggested that land surface plays the decisive role in the droughts that plague the West African Sahel. He noted the increase of surface albedo that results from the deterioration of vegetation, caused by overgrazing, and theorized that intensive desertification in the Sahel might have been responsible for the reduction of precipitation in the early 1970s. Charney's ideas have been partially reformulated by studies that show the drought to be part of a continental-scale rainfall anomaly (Figure 3.12). It is now believed that the onset

FIGURE 3.12 Map of African rainfall departures (percent below normal) for 1983, superimposed upon a Meteosat satellite image (any areas with rainfall exceeding the long-term mean are shaded). SOURCE: Reprinted, by permission, from Nicholson (1989). Copyright © 1989 by the American Geophysical Union.

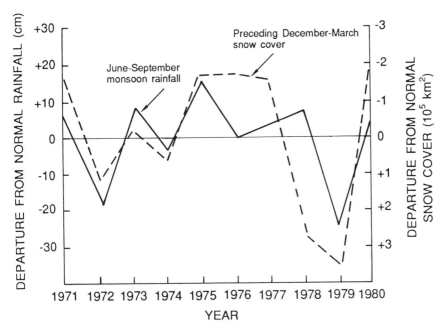

FIGURE 3.13 Indian summer monsoon rainfall and Himalayan snow cover of the preceding winter. Reprinted, by permission, from Walsh (1984) as adapted from Dey and Branu Kumar (1983). Copyright © 1983 by the American Geophysical Union.

of dry or wet conditions is initiated by large-scale atmospheric factors but that land surface processes reinforce them. Also, the emphasis is not now on albedo alone, but also on more comprehensive treatment of surface hydrology, which is presumed to play a role in maintaining prolonged drought patterns in areas like the African Sahel or the American Great Plains in the 1930s and in promoting the northward advancement of the Indian monsoon.

Most of the evidence for land surface influences on such large-scale climatic anomalies has come from numerical modeling studies. Thus far, the only strong observational support at the regional scale derives from studies of snow cover, an extreme case of land surface change. Namias (1978) argued that the persistence of certain weather patterns over the United States, such as the 1976-1977 winter with severe drought in the West and intense cold and frequent blizzards in the East, can be explained only by considering such factors as the extent of snow cover. Similarly, Eurasian snow cover appears to influence the Asian summer monsoon and other large-scale weather patterns with remarkable consistency (Figure 3.13).

GREENHOUSE GASES AND GLOBAL HYDROLOGY

The atmosphere contains various trace gases, both natural and man-made, that effectively trap the outgoing terrestrial radiation and warm the climate. Since the Industrial Revolution, concentrations of some of these so-called greenhouse gases, such as carbon dioxide, methane, nitrous oxide, and chlorofluorocarbons, have been increasing steadily in the atmosphere. If the present emission rates continue, the combined thermal forcing from atmospheric carbon dioxide and other greenhouse gases identified above could double from the preindustrial level sometime during the first half of the next century. It has been suggested that such increases of greenhouse gases may have a profound impact on our environment. Thus major effort has been devoted to the study of the climatic hydrologic change that may result from a future increase of greenhouse gases.

As early as 1938, Callendar suggested that the increase of atmospheric carbon dioxide due to fossil fuel combustion can warm the climate. His study and several others that followed are based on the radiative heat budget of the earth's surface and did not take into consideration other components of surface heat budget, such as the turbulent flux of heat and water vapor through the atmospheric boundary layer. It was noted that the carbon-dioxide-induced warming of the system is enhanced by an accompanying increase in the water vapor content of the air, which increases the infrared opacity of the atmosphere, thereby reducing outgoing terrestrial radiation. With the advance of computer technology, it has become feasible to use three-dimensional general circulation models (GCMs) of the atmosphere for the study of this problem. Accompanied by the general rise of surface temperature, the global mean rates of both evaporation and precipitation increase in the model, thereby intensifying the hydrologic cycle as a whole. The increase of downward terrestrial radiation due to the increase of carbon dioxide and atmospheric water vapor makes more energy available for evaporation. In addition, owing to the increase of the saturation vapor pressure at the surface, a larger fraction of energy is removed from the earth's surface through evaporation rather than through sensible heat flux.

The results from some numerical experiments also indicate that the future increase of greenhouse gases may affect regional as well as global hydrology. For example, owing to the penetration of warm, moisture-rich air into higher latitudes, precipitation and river runoff in the subarctic river basins are shown to increase markedly. Some of these experiments also predict that, in middle and high latitudes of the Northern Hemisphere, mid-continental soil wetness is reduced in summer due to the slight northward shift of the middle latitude rainbelt and the earlier termination of the snowmelt period in spring. On the other hand, winter soil wetness is found to increase in middle and high latitudes. The shift of the rainbelt also induces the reduction of winter precipitation in the subtropical steppe regions of the Northern Hemisphere.

The future hydrologic changes mentioned above are very broad scale phenomena and lack geographical details. It should also be noted that some of the changes do not represent the consensus from all numerical experiments, reflecting the unsatisfactory state of the art of current climate modeling. Therefore major effort is needed to improve various basic components of climate models, such as the parameterizations of moist convection, and the prognostic systems for convective and nonconvective cloud cover. In particular, more accurate macroscale determinations of evaporation, snowmelt, and runoff over each finite-difference grid box of a few hundred square kilometers are needed for the reliable prediction of future changes in climate and land surface condition.

One of the important factors that profoundly affect the long-term response of climate and hydrology to future increases of greenhouse gases is the thermal inertia of the oceans. To study this topic, it is therefore necessary to use a coupled ocean-atmosphere model in which a GCM of the atmosphere is coupled to an ocean GCM. Such a study of transient climatic response has just begun.

The prediction of climate change by the model should be validated against the observation of the actual climate change. Therefore in situ and satellite monitorings of the coupled ocean-atmosphere-land surface system and the factors causing global climate change are essential for this purpose.

Despite various uncertainties identified above, the results from current models suggest that the future changes in climate and the hydrologic cycle induced by greenhouse gases may be large enough to have far-reaching implications in agriculture and in the management of water resources. Therefore research into this topic should receive increased emphasis so that our adaptation to the future changes of climate and land surface conditions is facilitated.

The snow cover of the earth's surface is highly variable both seasonally and annually (Figure 3.14). Because the albedo of snow (the reflectivity over the entire solar spectrum) is very high compared to that of other common surface materials (Table 3.2), its presence or absence has an enormous influence on the earth's energy budget, both globally and regionally. A high albedo over a significant portion of a continent has a profound cooling effect on air masses in

What are the physical factors that control the snow cover-climate feedback process and its role as amplifier of climatic change?

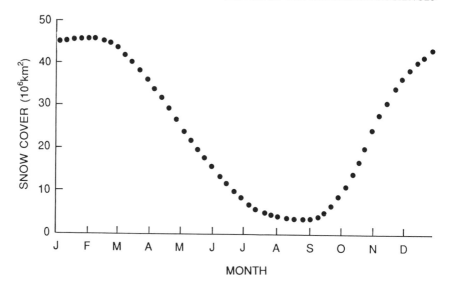

FIGURE 3.14 Areal coverage of Northern Hemisphere snow cover (10^6 km^2) from the climatology of K. F. Dewey and R. Heim (1981). SOURCE: Courtesy of the U.S. Department of Commerce.

high latitudes while reducing the total input of solar energy absorbed by the earth's surface. The cold air mass flows out of the Arctic and reaches and interacts with the surrounding, warmer water bodies. The cooling effect of snow cover also increases the summer-winter contrast of surface air temperature in high latitudes.

The interaction between snow cover and overlying atmosphere

TABLE 3.2 Reflectivity (in percent) of Various Surfaces in the Spectral Range of Solar Radiation

Bare soil	10-25
Sand, desert	25-40
Grass	15-25
Forest	10-20
Snow (clean, dry)	75-95
Snow (wet and/or dirty)	25-75
Sea surface (sun >25° above horizon)	<10
Sea surface (low sun angle)	10-70

SOURCE: Reprinted, by permission, from Kondratyev (1969). Copyright © 1969 by Academic Press.

constitutes a snow albedo feedback process that substantially affects the sensitivity, variability, and stability of climate. When the near-surface temperature of the atmosphere drops, a larger fraction of precipitation falls as snow and the melting of snow becomes less likely, resulting in expansion of the snow-covered areas. Because of the increased reflection of solar energy from the expanded snow cover, the sensible heat flux from the earth's surface to the atmosphere is reduced, further reducing the surface air temperature.

This positive feedback process amplifies the temperature anomalies in high latitudes. In addition, it contributes to the persistence of such an anomaly. The process is also responsible for the predicted polar amplification of the global warming induced by future increases of greenhouse gases.

Our concerns with snow albedo necessarily start at the scale of snow grains (about 1 mm) because the albedo of snow itself varies significantly with a variety of snow properties as well as the wavelength of the radiation (Figure 3.15). Snow properties change in time. First, the grains are always growing and change shape often, especially near the surface, where wind and other processes affect the snow in a continually changing sequence of events. Second, when wetted, snow assumes a noticeably darker appearance because of its reduced reflectivity.

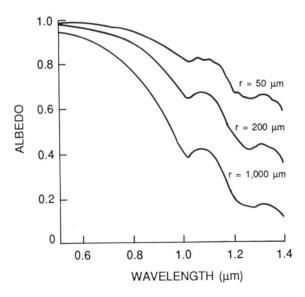

FIGURE 3.15 Albedo of snow of semiinfinite depth for various grain sizes and solar zenith angle of 600. SOURCE: Reprinted, by permission, from Dozier et al. (1981). Copyright © 1981 by the American Geophysical Union.

WHY IS SEA LEVEL RISING?

News stories about the greenhouse effect have awakened a new appreciation of how our environment is changing. One of the most serious of these changes is a projected rise of global sea level.

Will the rise, by the middle of the next century, be 10 cm, 1 m, 2 m, or, as some rather extreme scenarios have suggested, as much as 3 to 5 m? Imagine the impact of just a 1-m sea level rise, with the accompanying 100 m of shoreline retreat along flat-lying coastlines, on coastal communities, coastal hydrology and estuarine development, coastal wetlands, aquifers, and beaches.

Why will greenhouse-gas-induced climatic warming cause sea level rise? Two effects are involved: (1) the ocean volume will expand as it is heated by the warmer atmosphere, and (2) glaciers and ice sheets will melt, with ensuing runoff of meltwater to the sea. Of these two, meltwater is likely to be the more important factor in the long run—and this is basically a hydrologic problem.

Atmospheric circulation models predict that, with a doubled concentration of greenhouse gases in the atmosphere, the surface air temperature may increase by a few degrees centigrade on a global average, and the increase is likely to be greater at high latitudes. This would produce negligible surface melting in Antarctica and probably an increase in snow accumulation there. However, increased melt is highly likely in Greenland, the icecaps of the Canadian, Scandinavian, and Soviet Arctic, and the glacier-covered areas of Alaska, Central Asia, and the southern Andes.

We need good estimates of this increased meltwater production and runoff to the ocean to advise those who will have to study the scenario of an encroaching sea. But the problem is complex. Estimating the amount of increased snowmelt caused by warmer climate scenarios is relatively straightforward because snow surface energy-balance models are reasonably well developed. This additional snowmelt does not necessarily cause increased runoff, however. Over much of Greenland and the Arctic, the snow is so cold that meltwater refreezes within the snowpack and none escapes to the sea. With increased meltwater production, however, the refreezing of water in the snowpack warms the snow until finally it is brought up to the freezing point and water can then move through it and become runoff. This process is just now being investigated in detail, but many questions remain. It is likely that it will take decades to centuries to warm these vast, cold snow and ice masses to the point where additional runoff will be produced.

In addition to this problem, other interesting questions involving hydrologic processes remain. For instance, with a warmer climate, how will the rate of iceberg breakoff (calving) be affected? Iceberg calving in Alaska appears to depend on, in part, the meltwater runoff and the rate of stretching flow of the ice (which depends on the subglacial hydrologic system). Are these same processes important in, for instance, Greenland and Antarctica? If so, they would be critical to the stability of ice sheets such as the West Antarctic ice sheet, which is grounded well below sea level. Triggering an unstable disintegration of the West Antarctic ice sheet would cause a frightening sea level rise of 5 to 7 m, but recent studies suggest that this might take thousands of years.

Rising sea level requires responses by our planning, political, and scientific communities. Much of the economic and social cost can be reduced if adequate forewarning can be given. The important question for hydrologists and others in the geoscience community is whether the rate of sea level rise can be predicted with sufficient confidence, and in a sufficiently timely manner, that appropriate actions can be taken.

Third, the albedo of snow is very sensitive to contamination, and our planet is greatly contaminated by our activities. The underlying importance of this subject to large-scale processes affecting the entire planet has led to continuing investigations of this single property of snow (e.g., Warren, 1982). The basic information generated is needed to determine the albedo of snow in large-scale models of atmospheric circulation.

The measurement of albedo and many other important snow properties by large-scale remote sensing is a rapidly developing field (see Chapter 4). Remote sensing offers many possibilities for data collection that should be fully exploited to gather information about the snow cover over the large scale of interest for snow-climate studies. The use of remote sensing includes measurements of fundamental interest to snow hydrology such as snow-covered area and, more recently, the possibility of determining albedo directly. Remote sensing offers the advantage of averaging a snow property over large areas. The single most important thing to recognize about snow is that it is always changing, especially at the surface, where winds, solar input, temperature, moisture, condensation, and precipitation are always changing the textural characteristics of the upper layer. This greatly influences the albedo.

It is only in a very shallow layer at the surface that the albedo is determined because visible radiation does not penetrate snow to a great depth. Nevertheless, the depth of snow is a very important parameter in the determination of snow-climate feedback because the amount of snow that must be melted to reduce the albedo back to that of the underlying surface is an important consideration in determining climate variability. For example, there are some model indications

WARREN THORNTHWAITE
(1899-1963)

Warren Thornthwaite has been described by some peers as "the most influential climatologist of his generation." This praise was earned in part for his fundamental studies into the nature of evaporation and his formulation of the concept of potential evapotranspiration, for his outstanding 1948 paper "An Approach Toward a Rational Classification of Climate," for his establishment of the Laboratory of Climatology at Seabrook, New Jersey, which became a mecca for climatologists and hydrologists from all parts of the world, and for his early leadership of the World Meteorological Organization's Commission for Climatology.

Thornthwaite was born in Bay City, Michigan, and spent his early childhood helping on the family farm. He graduated from Central Michigan University (then a teachers' college) in 1922 and undertook some graduate work at the University of Michigan before moving on to the Department of Geography at the University of California, Berkeley, where he came under the influence of Carl Sauer and graduated with a Ph.D. in 1929. Although his dissertation involved an urban geography study of Louisville, Kentucky, his first academic appointment was at the University of Oklahoma (1927 to 1935), where, with the beginning of the Dust Bowl, he came face to face with the important role moisture played in human lives. Here his first two papers were published on a new system of climatic classification, superseding Koeppen's widely accepted classifications. These new classifications involved ideas of thermal and rainfall efficiency and used the rainfall-evaporation balance as the basic control of world soils and vegetation distribution.

From 1935 to 1946, Thornthwaite was chief of the Climatic and Physiographic Division of the Soil Conservation Service, U.S. Department of Agriculture, where he began both theoretical studies and field observations on evapotranspiration. This work led to many papers on the moisture factor in climate and to his development of the concept of potential evapotranspiration that became the basis of his famous 1948 climatic classification. Thornthwaite took leave from the government in 1946 to advise Seabrook Farms, a large grower and packager of frozen vegetables, on irrigation. While there, he was asked by the Air Force to undertake basic micrometeorological studies. He established the Johns Hopkins University Laboratory of Climatology in Seabrook (later moved to Centerton, New Jersey). Under his insightful guidance, outstanding contributions were achieved in basic and applied climatology, in micrometeorological instrumentation, and in the use of spray irrigation to purify food-processing effluents. The reputation of the laboratory was such that leading climatologists and hydrologists from all parts of the world visited to confer with the research staff.

Warren Thornthwaite was a complex man. His writing was simple and clear. He was not receptive to criticism, but, at times, he could be critical of others. Few of his peers could be entirely neutral about Thornthwaite, but those who knew him well recognized the deep-seated dedication and love that he had for his chosen field of work and for his fellow man.

that increases in carbon dioxide could cause soil desiccation in some of our prime agricultural areas because of the early meltout of the snow cover in the spring. In general, any climatic change in the temperature of polar regions will directly involve the seasonal snow cover or permanent snows of the polar ice sheets.

The climate-snow cover feedback involves not just albedo but also the effects of temperature and airborne particles, including aerosols. For example, the fact that the albedo of snow is so sensitive to contamination implies that the scenario of "nuclear winter" (the climatic response to the aerosols and dust injected into the atmosphere following a major nuclear exchange) would be sensitive to the wide range of albedos that a typical snow cover experiences, from fresh snow to slightly dusted snow. The temperature-snow cover feedback must include the reduction in albedo associated with the onset of melt in snow.

To understand the climate or even short-term weather patterns of a snow-covered continent, it is necessary to understand a wide range of the physical properties of snow and to use these to determine the boundary conditions for models of atmospheric circulation.

HYDROLOGY AND WEATHER PROCESSES

Introduction

Of all the processes in the hydrologic cycle, precipitation in all its forms probably receives the most public attention. It is easy to grasp its importance in replenishing surface and subsurface flows and water levels; floods and droughts affect our lives and livelihood in no small measure. Weather processes include all the atmospheric processes that produce precipitation, from the microphysics of cloud and precipitation particle growth to the continental and global patterns of airflow that control the behavior of weather systems.

Because precipitation is a product of this complex combination of dynamic, thermodynamic, and cloud microphysical processes, it is among the most difficult meteorological quantities to model and forecast. These processes operate over a variety of space and time scales, which exhibit control and feedback mechanisms among themselves and also interact with surface orography, roughness discontinuities, and soil moisture. The nonlinearity in the governing equations makes the predictability problem a major scientific challenge, as demonstrated in the now-classic works of Edward Lorenz on chaos, a topic further discussed in the last section of this chapter, "Hydrology and Applied Mathematics." Therefore understanding the interactions between hydrology, weather, and climate and applying this knowledge to solving real-world problems point to research frontiers that go well beyond the traditional concerns of these disciplines.

A ubiquitous feature of rainfall—ranging in time from a few minutes and hours to months, years, decades, and even centuries, and in space from a few to several thousand square kilometers—is the presence of extreme variability. There is now widespread interest among meteorologists, climatologists, and hydrologists in measuring, modeling, and predicting the nature of this variability, using a variety of new mathematical tools and observational capabilities from land and space. The emphasis in meteorology is largely on the physics of rain formation, the dynamics governing the temporal and spatial rainfall distributions, and short-term forecasting. In climatology, dealing with the problems of global climate, precipitation is recognized to be an important source of energetics both in tropical regions and in extratropical climates. For example, efforts are under way to measure global rainfall using satellites. In hydrology, the interest is in its effect on streamflow, on soil moisture, and on understanding and predicting the response of river basins and aquifers to climatic fluctuations spanning a broad range of space and time scales.

Evaporation and transpiration from the continents affect the weather. In the tropical rain forests of the Amazon there is a high degree of recycling between evapotranspiration and rainfall. On the other hand, in desert latitudes the absence of significant surface water vapor sources may have some influence in maintenance of the desert itself. **Human-induced changes in land surface characteristics are accelerating around the globe, and thus it is urgent to understand the influences of land surface processes on weather and climate.** Once again, we stand on the threshold of a time in which a new array of tools, notably remote sensing of the land surface from satellites, in combination with targeted field experiments, promises advances on this challenging and important set of problems.

Some Frontier Topics

Land Surface-Atmosphere Interaction

What are the reciprocal influences at the mesoscale between land surface processes and regional weather?

What happens to the surface fluxes of water and energy when an area of natural grassland, 100 × 100 km, is plowed and converted to irrigated agriculture? When a comparable area of tropical rain forest is cut down and converted to agriculture? When a rainstorm soaks a comparable area of the midwestern United States that had been parched by drought? Do the resultant changes have an impact on weather?

Statistical studies have established the role of land surface characteristics in determining mesoscale meteorological patterns and in the development of severe weather events. Irrigation in the Great Plains appears to have a significant influence on convective activity, enhancing rainfall and locally increasing the frequency of thunderstorms and tornados during the warm season. Thus a better understanding of land surface-atmosphere interaction may lead to improved forecasting of severe weather events.

This aspect of land-atmosphere interaction is particularly exciting because it can benefit from advances in the rapidly developing research area of mesoscale meteorology. Historically, meteorology has emphasized the larger scales of planetary waves, cyclones, and anticyclones (synoptic meteorology), and also micrometeorology, describing turbulent transfer in the surface and planetary boundary layers. For decades, the scales in between (literally how the mesoscale was defined) were thought to be scales where not much was happening, and this supposed "spectral gap" permitted turbulent and convective processes to be parameterized in models of the synoptic scale. However, the atmospheric mesoscales include such energetic phenomena as thunderstorms, squall lines, and other mesoscale convective systems, important in their own right and not yet treated properly in larger-scale models, including GCMs of weather and climate.

Progress in mesoscale research has long been impeded by a lack of observational capabilities, techniques, and models applicable to spatial scales of kilometers to hundreds of kilometers. This situation is changing rapidly. Satellites are providing much of the needed data, and appropriate ground-based instrument systems will soon provide mesoscale information from which detailed energy and water budgets can be derived. For example, in the early 1990s a network of 30 surface-based radar wind profilers will be deployed on spatial scales of 200 to 400 km over a large area of the central United States. A next-generation weather radar (NEXRAD) network of Doppler radars (see Chapter 4) will provide greatly improved precipitation and boundary layer wind estimates over most of the United States by the mid-1990s. Because these networks will be operational, data potentially can be made available at low cost, and the research community can concentrate resources where they are needed the most; i.e., it can supplement the operational data with targeted campaigns to add high-quality research data in specific test regions. This will be the strategy of the National STORM Program (see Chapter 4), a decade-long partnership between research and operational atmospheric scientists in the 1990s. These test regions can be used, in turn, to validate and improve remote sensing methodologies for use on next-generation satellite systems.

Appropriate techniques for bridging the various spatial scales from microscale to mesoscale to synoptic are also emerging, in part because of a number of field programs aimed at improving the compatibility of field and satellite observations with the needs of hydrologic and atmospheric models. These include the HAPEX project in France and the First ISLSCP Field Experiment (FIFE) in the Great Plains, both described in Chapter 4. State-of-the-art numerical models of the atmospheric general circulation, necessary for climate studies, use horizontal grid dimensions of the order of 100 km and may soon be in the 50-km range, but not much less. These are also spatial scales characteristic of important catchments. Thus it is imperative that we improve techniques for diagnostic studies on these scales.

One of the crucial terms in atmospheric water budgets, the horizontal flux divergence, can be measured with reasonable accuracy on a 500-km scale, but the error is inversely proportional to the length scale, and experimental estimates are very poor on scales of 100 km and less. Hydrologists who study processes that govern the surface water balance, and hence evapotranspiration, must consider variations in the properties of soils and vegetation that are important on scales of a few meters, and there are no accepted methods for estimates on kilometer and greater scales. Multidisciplinary studies are under way that attempt to bridge this scale gap by increasing understanding of the fundamental hydrologic processes that can lead to parametric formulations for land surface processes on the 10- to 100-km scale.

Under what conditions does the spatial distribution of evaporation generate regional circulations that have a marked influence on mesoscale rainfall?

For a long time, laypersons and scientists alike have assumed that simply increasing the rate of evaporation will increase precipitation. However, recent research suggests that at the mesoscale it is the spatial distribution of evaporation, rather than its absolute magnitude, that influences precipitation. The largest factor, it appears, is circulation generated by differential heating between wet and dry areas. To show that this is a plausible hypothesis, consider how little rain falls over much of the subtropical oceans, where evaporation is very great. Yet along coastlines, where a greater fraction of solar energy goes into sensible heating over land (and less into evaporation), such as Florida's, a temperature gradient develops and generates a circulation, well known as a sea breeze, which in turn generates thunderstorms almost daily. It remains to be seen whether boundaries between wet and dry areas (for example, between irrigated and nonirrigated land) can generate analogous inland sea breezes strong enough to affect precipitation patterns. Testing such hypotheses will require field data to establish water budgets over 100-km-scale regions, and the new operational networks are an important step toward this capability.

Before we can answer such questions about how alterations in land surface properties can alter weather and climate, and improve predictions of elements of the hydrologic cycle regionally, we must learn how to make quantitative estimates of regional water budgets on both the atmospheric and the terrestrial sides of the interface. We want to know the daily and seasonal cycles of these fluxes and storage terms and their sensitivity to soil and vegetation type, state of growth, and precipitation history, including the effects of uneven distribution of precipitation.

Models for Mesoscale Convective Systems and Applications to Flash Flooding

---

What kind of weather system produces excessive rainfall and flash flooding?

FLASH FLOODS

Flood hazards in small drainages are increasingly affecting the safety and economic well-being of the citizens of the United States. The loss of life on rapidly cresting rivers is now the major weather-related killer in the United States, and the annual loss of property associated with floods is approaching $2 billion. Some recent examples follow:

• August 1955—northeastern United States—184 deaths and flood damage totalling over $1 billion from widespread floods resulting from the remnants of former Hurricane Diane; 20 inches of rain fell in the region from Pennsylvania to New England.

• August 1969—James River Basin, Virginia—153 deaths and more than $100 million in property damage as dying Hurricane Camille dumped almost 30 inches of rain in less than 8 hours.

• February 1972—Buffalo Creek, West Virginia—118 deaths and hundreds of homes washed away as a dam made of coal mine waste gave way after heavy rains.

• June 1972—Rapid City, South Dakota—236 deaths and $100 million in property damage after a nearly stationary mesoscale convective system unleashed 12 inches of rain in a few hours on the east slopes of the Black Hills.

• June 1972—northeastern United States—120 deaths and about $4 billion in property damage as the remnants of Hurricane Agnes produced widespread and destructive flooding and flash flooding from Virginia to New York; 14 inches of rain fell in 24 hours.

• July 1976—Big Thompson Canyon, Colorado—139 deaths and more than $50 million in property damage after a nearly stationary mesoscale convective system over the east slopes of the Rockies deluged the western third of the canyon with 12 inches of rain in less than 6 hours.

• July 1977—Johnstown, Pennsylvania—76 deaths and more than $200 million in property damage when a nearly stationary mesoscale convective system over central Pennsylvania deposited up to 11 inches of rain on a seven-county area in 9 hours.

• September 1977—Kansas City, Missouri—25 deaths and $90 million in damage when a nearly stationary mesoscale convective system deposited 11 inches of rain in a few hours, turning "gentle" Brush Creek, which flows through the heart of the city, into a raging torrent.

• November 1977—Taccoa, Georgia—40 people, half of them children, died when heavy rains ruptured an earthen dam and demolished a mobile home community in the valley below.

• August 1978—central Texas—33 deaths and tens of millions of dollars in damage as very heavy rains associated with the remnants of Tropical Storm Amelia fell over the area.

• February 1980—southwestern United States—22 deaths and more than $300 million in property damage as heavy rains caused floods, flash floods, and mudslides.

• October 1981—northern Texas and southern Oklahoma—5 deaths and damage approaching $200 million as a nontropical system combined with decaying Tropical Storm Norma to produce very heavy night-time thunderstorms.

• January 1982—west-central California—more then 30 deaths and almost $300 million in damage due to heavy rains, flooding, and mudslides.

Source: STORM DATA, published monthly by the National Climatic Data Center. National Environmental Satellite, Data, and Information Series. National Oceanic and Atmospheric Administration, Asheville, N.C.

Although several plausible relationships between weather systems and extensive precipitation exist, such as large-scale cyclones and their associated fronts and tropical cyclones, the size and strength of large-scale disturbances are at best poorly correlated with rainfall intensity.

For example, a meter of rain fell in four days in Manila from a barely detectable wind disturbance. The Rapid City, Big Thompson, Kansas City, and Johnstown flood disasters are all examples lacking a significant cyclonic storm or strong front. The apparent paradox extends to tropical cyclones; many powerful hurricanes produce moderate rainfall, whereas some of the worst flood disasters are associated with weak or dissipating storms.

Failing an explanation in the large scale, we turn to the scale of the building block of many rain events, that of the thunderstorm. Our basic knowledge of the individual thunderstorm owes much to the classic 1946-1947 field program known as the Thunderstorm Project (Byers and Braham, 1949). There are two basic difficulties in attributing flood-producing rains to the thunderstorm scale. First, the individual storm lives only about an hour, and its intense rain may cover a swath just several kilometers wide. Second, there is little evidence to support a significant correlation between thunderstorm severity and rain volume. Examples abound of violent storms producing hail, winds, and even tornados without flooding, whereas, paradoxically, few flood disasters (including those cited above) are accompanied by thunderstorms of exceptional severity.

Meteorological analysis of flood-producing rain events themselves has given us some important pieces of the puzzle. The spatial scale of these events ranges from tens to a few hundred kilometers, and the temporal scale is several hours, not minutes or days. This is firmly within the atmospheric mesoscales, literally intermediate between the typical cyclonic and thunderstorm scales. Knowledge of mesoscale processes has advanced greatly during the last two decades, aided by the application of satellite remote sensing, radars, and in situ measurements on the mesoscale.

The essential new finding is that large-scale disturbances do not merely produce scattered "popcorn" convective clouds and storms that remain closely coupled to the large-scale controls. Often, the convective clouds quickly become organized into obvious mesoscale structures that can move independently of and occasionally away from the large-scale disturbance. Certain mesoscale systems resemble two-dimensional lines of convective storms that are called squall lines. Others develop a nearly circular cloud shield as viewed by satellite and are called mesoscale convective complexes. The generic term "mesoscale convective system" (MCS) is used to refer to most convection organized on the mesoscale.

The conceptual model that describes the essential features of most MCSs and their life cycle is shown in Figure 3.16. The deep convective clouds form and quickly become organized, often into one or more lines. The convective clouds continue to grow and die along this same line for many hours; the line may be slow or fast moving and may propagate with a velocity different from that of the ambient wind at any level. After some hours, the stratiform precipitation region forms, its water source a combination of water vapor and ice crystals advected from the debris of the continually dying convective clouds in the adjacent line, and independent mesoscale ascent.

There is a formidable challenge to theoreticians to explain circulations on the scale of MCSs. There appear to be no fundamental modes of atmospheric waves on this scale. Convective clouds scale with the depth of the troposphere (10 to 15 km), whereas consideration of the effects of condensation heating in a stratified rotating fluid suggests circulations on the order of 1,000 km in horizontal scale. Only in the past several years have theories been proposed, and only for the squall line type of MCS. Theories and numerical models emphasize the importance of low-level vertical wind shear and of the cold pool of air produced by the organized convective downdrafts.

The ubiquity of MCSs represents an outstanding example of how new observations in a geophysical science, in this case from satellite and radars, led the science in new directions. Not only was there no place in accepted theory for such mesoscale weather systems, but they are also a major inconvenience for atmospheric modelers, whose lives would be easier if the mesoscale spectral gap were free of energy-producing systems. As Palmer (1952) so well put it, ". . . it is not only that the griffins and basilisks described by the philosophers are

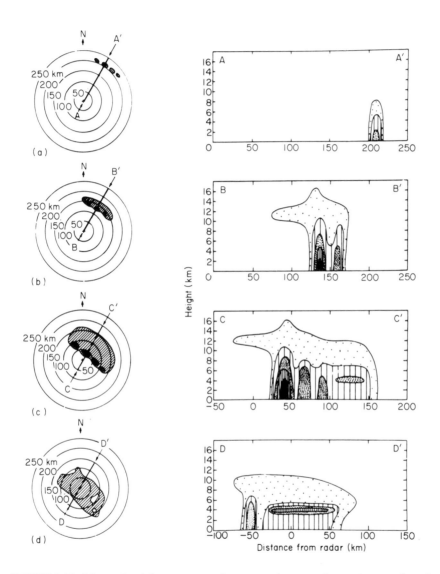

FIGURE 3.16 Schematic of the structure of a mesoscale convective system as viewed by radar in horizontal and vertical cross sections during (a) formative, (b) intensifying, (c) mature, and (d) decaying stages of its life cycle. SOURCE: Adapted, by permission, from Leary and Houze (1979). Copyright © 1979 by the American Meteorological Society.

absent; it seems that the country is occupied by creatures of which they have never dreamt."

Descriptive models of a particular kind of MCS, the mid-latitude prefrontal squall line, were developed from data obtained during the Thunderstorm Project (Newton, 1950). Newton also pointed out the likely role of the cold air outflows in continuous triggering of new storms and therefore in the propagation of the system. It took more than two decades for researchers, stimulated by tropical field experiments, to point out the essential similarities among MCSs around the world. Within the last 15 years, many investigators have developed conceptual models of the structure, propagation, and life cycle of MCSs, including the central role of evaporation of precipitation in producing downdrafts and how they aid the upscale development of these systems.

We are now in a position to describe the phenomena of most flood-producing rainstorms. The basic ingredient is an MCS, with the addition of a fairly restrictive condition: the convective portion of the MCS must remain nearly stationary for a period of a few hours. Basically, many individual convective storms must form and move along the same track. This amounts to a requirement that the propagation vector of the convective system be opposite to the vector describing the motion of individual cells (Figure 3.17).

It is quite possible that just the right thermodynamic and wind shear environment exists to produce a stationary MCS. More frequently, the environment favors an MCS that would move very slowly, and some focusing mechanism "locks in" the process over a particular region. This can often be the edge of the cold pool from a previous MCS; over flat terrain this may be the frequent cause of flash flooding, as, for example, in the Kansas City event. In many cases, however, the lifting associated with airflow interacting with topography can anchor the convective region; notable examples are the Rapid City and Big Thompson disasters.

New mesoscale data sets of unprecedented quantity and quality will be available to both researchers and forecasters during the next several years. Numerical models have now attained the sophistication required for simulation of MCSs, and they will assist in developing a theoretical basis for understanding MCSs of various types and their evolution and interaction with their environment. There is an important additional factor that must be supplied by hydrologists: what will be the response of the river basin? Improved warnings of flash floods require that better rainfall observations and predictions be coupled with hydrologic models of runoff production on the basin scale.

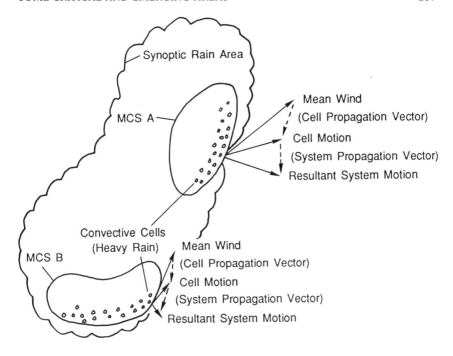

FIGURE 3.17 Schematic mechanism for flood-producing rains. MCS A moves rapidly and produces a few brief heavy rain bursts as its convective cell region passes any given location. MCS B is the same size, shape, and intensity as MCS A but produces a very different result. Its system propagation vector is nearly equal and opposite to its cell motion vector, giving a resultant system motion so slow that many convective cells will form and drop heavy rain over the same limited area. SOURCE: Adapted from Chappell (1987).

Stochastic Modeling of Space-Time Variability in Rainfall

What is the adequate statistical structure to represent the organization and evolution of mesoscale storm systems in space and time?

The scales of the physical and dynamical processes that produce and distribute precipitation constitute a continuum. Therefore for both theoretical and practical reasons a model describing or predicting precipitation can have only limited temporal and spatial resolution.

Atmospheric processes at the unresolved scales have to be treated statistically because the information needed to describe these processes is not available or because the physical processes themselves lead to inherent fluctuations at such scales. In this latter case, owing to the nonlinear and turbulent nature of atmospheric flow, even processes resolved by a model have only limited ranges of predictability in the deterministic sense. (For further discussion see "Hydrology and Applied Mathematics," the last section of this chapter.) Beyond the limits of predictability, only statistical treatment of the processes is possible. A major task in precipitation modeling is to determine the statistical structure of the unresolved processes and to couple this statistical structure to the physics and dynamics of the processes at the resolved scales.

Much of the recent emphasis in precipitation modeling in hydrology has been on identifying an appropriate class of stochastic models to conceptually represent the organization and evolution of mesoscale storm systems. Typically, the building blocks of these models constitute the life cycle and motion of a convective rainfall cell, and the spatial organization of cells as groups of clusters within a moving mesoscale storm system. The cells can move with the same velocity as the mesoscale system or with a different velocity relative to it. The rainfall intensity in time and space is spread deterministically around the center of each cell. The rainfall intensity at the ground level is then represented as the sum of rainfall contributions from the groups of cell clusters within a moving mesoscale storm system.

Various statistical assumptions are introduced on these components of a storm system so that the statistical features of the resultant ground-level rainfall intensity can be evaluated analytically or numerically and tested against rainfall observations. Typically, these features include space-time correlations, probabilities of extreme rainfall, and crossing properties with respect to different thresholds. The available empirical information regarding mean intensity at the center of the cells, cell duration, birthrates, spatial extents, and mean number of cells in a cluster allows for incorporation of these values as parameters in stochastic models. Moreover, a stochastic model also allows for the incorporation of variability in these entities. Such variability is ubiquitous among different storms in a given climate and among storms in different climates. Figure 3.18 shows an example of a stochastic simulation of air mass thunderstorm rainfall.

Although the line of research explained above was initiated about three decades ago, not much progress has been made in this realm owing to mathematical difficulties inherent in analyzing the space-time stochastic structures described above. However, recent research shows the promise of statistical models in providing a physical understanding of mesoscale precipitation over a broad range of space and time scales. For example, Waymire et al. (1984) derived an analytical expression of space-time rainfall correlation from a stochastic model that shows that the pattern of temporal fluctuations propagates in space as a frozen field through the velocity of the storm system. This is known as Taylor's hypothesis in fluid turbulence. After time intervals of the order of the mean lifetime of a convective cell, the storm reorganizes itself, and a new pattern of temporal fluctuations appears. Therefore the range of validity of Taylor's hypothesis in space-time rainfall seems to be controlled by the mean life span of a convective rainfall cell. In this sense, stochastic models are helping to unify disjoint empirical observations of rainfall, such as the observed mesoscale organization of storms, on the one hand, and the observed correlation structure of space-time rainfall, on the other.

One of the major thrusts of future research in stochastic modeling concerns the testability of conceptual mathematical constructs, as described above, against empirical observations of rainfall. These observations typically are available only as averages in time over different intervals such as hours, days, months, and years and in space over regions of different areas. Stochastic models of temporal rainfall are being tested by analytically computing statistical properties of the integrated rainfall over different time intervals (e.g., extremes, proportion of wet/dry periods) and testing these predictions against empirical observations. The goodness of the mathematical construct is judged by its ability to predict the statistical characteristics of the aggregated rainfall over different intervals that are not used a priori in specifying the model parameters. Indeed, the stochastic models exhibiting clustering in time, as suggested by the space-time constructs, have shown the most promise in this realm.

How can the necessary and fundamental links between the deterministic (dynamic) and stochastic models of rainfall fields be established?

Preliminary analysis of spatial rainfall data over different-sized regions shows the presence of fundamental structures that are likely to be of great theoretical and practical importance. For example, the variance of area-averaged rainfall from the Global Atlantic Tropical Experiment (GATE) of the Global Atmospheric Research Program is observed to decrease very nearly as $A^{-1/3}$ with area A, as A ranges from

16 to 40,000 km². This phenomenon suggests that spatial rainfall may not have a unique length scale, for otherwise the variance would decrease as A^{-1}. Such processes are statistically self-similar, or scaling, insofar as the large-scale fluctuations remain statistically similar to the small-scale fluctuations. Simple stochastic models recently have been proposed that exhibit these types of scaling relationships in spatial rainfall. Most of these models consider spatial rainfall or temporal rainfall at a fixed point, and extensions to space-time are introduced by appealing to Taylor's hypothesis of fluid turbulence. The empirical basis for Taylor's hypothesis and the range of its validity

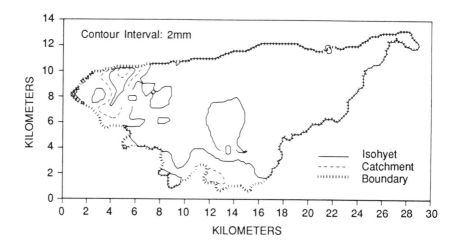

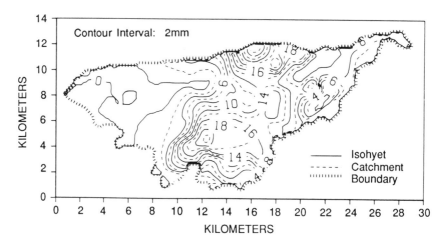

in diverse rainfall systems still remain to be established. The short lifetime of intense rainstorms argues against universal validity. Therefore a major goal of research in stochastic modeling is to characterize space-time intermittency or spottiness in rainfall as it pertains to various notions of scaling as well as the physically observed features of clustering, growth, and decay of convective cells, and larger-scale spatial forms observed in mesoscale rainfall systems. The intermittency problem has a long and rich history in the stochastic-dynamic theories of fully developed turbulence and is still a very active area of scientific research. Further discussion of this topic appears in the last section of this chapter.

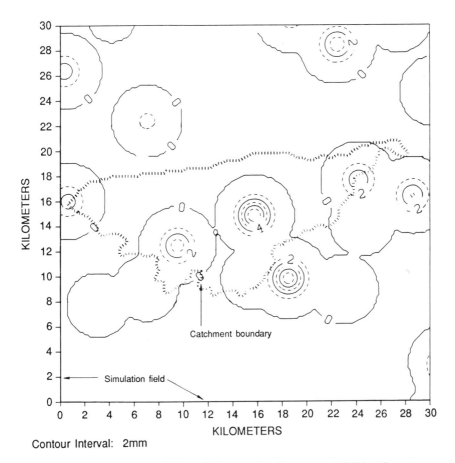

Contour Interval: 2mm

FIGURE 3.18 Stochastic simulation of air mass thunderstorm rainfall (catchment area = 154 km^2). SOURCE: Reprinted, by permission, from Eagleson et al. (1987). Copyright © 1987 by the American Geophysical Union.

The dichotomy in the current state of the art between the stochastic approaches and the deterministic approaches is unsatisfactory. It is important to understand the physical and analytical connections between these two types of approaches in modeling rainfall fields. For example, the dynamics involved in the formation and maintenance of mesoscale rain bands are not well understood, and several theories have been proposed. These theories generally appeal to the growth of one of several kinds of instabilities known to exist in the solution of the nonlinear dynamical equations governing atmospheric behavior. We can be optimistic that the combination and improvements in both the dynamical models and the data sets will lead to the rejection of some theories and the validation of others. We will then be in a better position to make the connections between the dynamical and the stochastic descriptions of rainfall, to determine their range of validity, and to obtain parameterizations of stochastic models in terms of measurable physical quantities.

HYDROLOGY AND SURFICIAL PROCESSES

Introduction

Surficial processes include the transport of mass and energy at and through the interface between the lower atmosphere and the earth's surface. As precipitation falls over land, part of it is intercepted by vegetation and other surface structures, and part of it reaches the ground, where it may infiltrate into the soil or run off directly over the surface into the nearest stream. Where precipitation falls in the form of snow, these same processes occur, but they are delayed until melting occurs. Over ocean and sea surfaces the precipitation reaches the water body directly. Between rainfall events, there is a continuous return flux of the water available at the surface to the atmosphere in the form of evaporation. Clearly, mass transport of water is the common denominator of these surficial processes in the hydrologic cycle. In addition, processes such as evaporation, dewfall, snowmelt, and hoar frost formation involve the redistribution of large amounts of latent energy under nearly isothermal conditions, so that they profoundly affect the near-surface environment. Thus **it is useful to consider these surficial hydrologic processes not only as mass transport phenomena, but also within the context of energy transport.** In fact, it is almost impossible to do otherwise.

Precipitation, as the primary source of water, and solar radiation, as the primary source of energy, are the external forcing agents for all surficial processes. However, on land the state of the surface generally dictates the disposal and the distribution of precipitated water among the main surficial processes, such as infiltration and soil moisture flow, evaporation, and runoff. Again this state is characterized from two viewpoints: the surface budgets of water itself and of energy. Thus, as long as the surface soils remain partly saturated, rainwater and snowmelt water infiltrate into the permeable soil layers to join fully saturated flow regions. Wherever such zones of saturation occur at the ground surface, there is surface runoff. This runoff may initially take place as a thin sheet flow, but local irregularities soon cause the flow to gather in small gullies. This flow in turn collects to form rivulets and small streams. In areas where soils are deep and permeable, the water remains underground on its way to the stream channels. The combined availability of moisture and energy near the surface controls the return of surface water to the atmosphere through evaporation. In colder areas, these phenomena involve further interactions and phase changes between water in the frozen, liquid, and vaporous state.

The state of the art of accounting for and describing surficial processes holds both formidable challenges and unique opportunities. One theme common in the hydrologic literature today is that it will be possible to describe the global hydrologic cycle only when we understand the relationship among relevant hydrologic processes on different temporal and spatial scales. This is certainly the case for the hydrologic processes taking place at the surface. Indeed, **one of the main obstacles in understanding surficial processes is the high spatial variability of surface features and hydrologic variables.**

Over the past decades, we have made great progress in the development of appropriate formulations of hydrologic processes at various scales. However, **the issue of the linkage and integration of formulations at different scales has not been addressed adequately. Doing so remains one of the outstanding challenges in the field of surficial processes.** Recent advances in remote sensing from aircraft and satellites promise to shed light on these problems, however. Satellites, in particular, offer a unique opportunity to obtain homogeneous data sets relevant to surficial processes. When used in conjunction with surface observations, satellites are the ideal tool for investigations covering a broad spectrum of temporal and spatial scales. Conversely, however, it will be necessary to improve our present understanding of the physics of surficial processes if satellite technology is to reach its full potential.

Some Frontier Topics

*Characterization of Spatial Variability of Soil Properties
and Its Relation to Infiltration*

How can local observations of infiltration and soil moisture be translated to larger regions?

Can the impacts of chemical and microbiological processes on unsaturated soil properties be quantified?

The retention of soil moisture and the attendant runoff from naturally occurring rainfall, snowmelt, or irrigation are fundamental processes upon which civilization depends for food production, potable water, and navigable streams and waterways. Infiltration—the process that partitions precipitation or irrigation into that part temporarily stored in the soil and that part remaining on or flowing over the soil surface—depends on the soil, its relief, and its management. Early research on infiltration is associated with the names of Green and Ampt, Kostiakov, and Horton. Most of our present understanding of infiltration stems from theoretical investigations made in the laboratory and on 1-m^2 field plots isolated from many of the factors that are relevant in natural and large-scale environments. The contemporary, scientific challenge is to extend this small-scale understanding to larger domains—watersheds of different climatic regions, irrigated and rainfed agricultural and silvicultural areas, and intensively managed fields smaller than 1 km^2.

The transition from point observations and models to field or regional scales is extremely difficult because soils vary from location to location. They are a product of the processes of soil genesis, and they continuously change in time and space over the earth's surface. Over 10,000 soil series have been identified in the United States alone. The degree of this heterogeneity is heightened because of the unique behavior of one physical property—unsaturated hydraulic conductivity. Hydraulic conductivity is a parameter in the equation first proposed by Darcy in the middle of the nineteenth century to describe the flow of water in soils. However, as pointed out by Buckingham early in this century, this conductivity depends on the water content of the soil. In fact, its value for any given soil can decrease 10 million times as the soil dries from water saturation to complete dryness.

This unique property is a benefit in that it accounts for a soil's ability to absorb, retain, or transmit water under a variety of initial and local conditions. Yet it markedly exacerbates the heterogeneity problems, as illustrated in Figure 3.19, where the unsaturated hydraulic conductivity measured at two different locations separated by a distance of only 50 m is plotted against soil water content. Random variations of field-measured soil water content typically exhibit a range of ± 0.05 cm^3/cm^3 and may be related to each other over distances on the order of only 10 m. Note that the two curves differ by nearly 2 orders of magnitude for the same water content. Such variability forces estimates of hydraulic conductivity even at short distances to have an uncertainty of a hundredfold.

Deterministic models of infiltration and unsaturated soil water flow are giving way to spatially stochastic concepts. Interest now centers

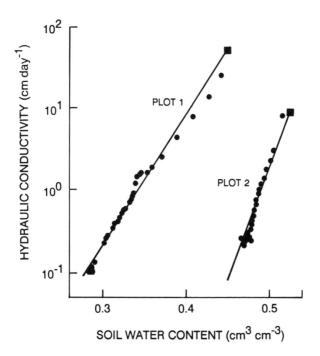

FIGURE 3.19 Illustration of the spatial variability that exists within a small region of a naturally occurring field soil. SOURCE: Reprinted, by permission, from Nielsen et al. (1973). Copyright © 1973 by the Regents of the University of California.

on how to describe the random distribution functions of unsaturated hydraulic conductivity and related soil parameters and the extent of their spatial correlation at various scales. This interest also has intensified the search for improved methods to collect and interpret sufficient quantities of data to ascertain the spatial variability within each soil series or soil-mapping unit. A conceptual framework defining the expectation and variance structure of the spatial and temporal heterogeneity of soil-water properties from local to global scales is greatly needed.

Some current research is unraveling the interactive physical mysteries of infiltration in field soils, but many additional challenges remain, particularly those that embrace chemically and microbiologically induced alterations of the energy status of water in the unsaturated soil. More than 80 years ago, Edgar Buckingham defined the concept of the "capillary potential" for unsaturated soils (Buckingham, 1907). No one has yet measured the water content or its potential energy within a single soil pore. The possibility of measurement becomes

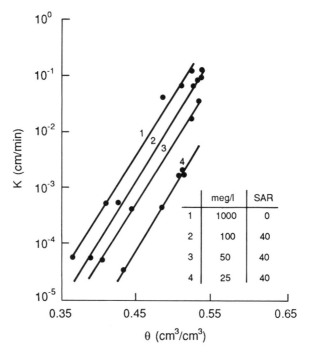

FIGURE 3.20 The impact of water quality (θ) on the value of the unsaturated hydraulic conductivity (κ). SOURCE: Reprinted, by permission, from Dane and Klute (1977). Copyright © 1977 by the Soil Science Society of America, Inc.

greater as tomography and differential tomography using gamma- and X-ray radiation, nuclear magnetic resonance, and other energy sources are applied to soil materials. These measurements will allow a microscale examination of the physics of water flow in films along soil particle surfaces, and, it is hoped, will reveal how the concentration and distribution of chemical constituents in the soil solution alter the macroscopically measured hydraulic conductivity. Figure 3.20 shows that the value of the hydraulic conductivity is drastically altered by the quality of the soil solution. Quantifying the many coupled and time-dependent microbiological-chemical-physical processes at the pore scale into a unified approach is an unmet challenge.

At the pore or laboratory scale, a number of issues remain unresolved. These include the hysteretic nature of the unsaturated hydraulic conductivity and soil-water characteristic curve, the effects of temperature on the hydraulic properties, the displacement of air during infiltration or drying, the simultaneous movement of water vapor and transport of heat, the impact of localized macropore geometries on leaching efficiency during infiltration, and the quantification of soil-water properties for soils that shrink and swell or crack and consolidate upon drying and wetting. Infiltration of organic liquids into moist soils is another issue. It is complicated by volatilization and the potential presence of co-solvents. Indeed, multiphase flow—including the migration of partially miscible and immiscible fluids—offers a challenge to the hydrologic community that only recently emerged, albeit in the form of a practical problem resulting from ground water contamination.

Runoff Production by Precipitation

One of the major difficulties in understanding and quantifying runoff generation in river basins stems from the presence of spatial variabilities in topography, geology, soil type, and vegetation, as well as in climate fluxes such as rainfall, infiltration, and evapotranspiration. Because of these spatial variabilities, even in a single basin, the spatial structure of runoff varies greatly from one rainfall event to another. Consequently, each rainfall event frequently contributes to only a part of the channel network system in a river basin, and this part changes from one rainfall event to another. Even though only a few experimental studies have attempted to monitor the time-varying nature of a channel network in runoff generation, the occurrence of this phenomenon must be widespread because it reflects the effect of spatial variability in runoff generation in basins. A central problem in river basin hydrology is to understand how river runoff is organized spatially, which also includes the issue of intermittency or spottiness.

Traditionally, the development of quantitative estimates of the transformation of precipitation to runoff has resulted primarily from the need to answer design and management questions related to civil works, and agricultural, rangeland, and silvicultural practices, and to forecast shortages and excesses of water at particular locations and over large geographical regions. Pioneering work in determining the mechanisms of flow production from catchments when the rainfall intensity exceeds the rate at which the rain could infiltrate the soil was done by Robert Horton between 1933 and 1945. At about the same time, Hursh and Brater (1941) demonstrated that in steep, forested landscapes significant volumes of stormflow could enter channels by subsurface paths. In the 1960s, it was shown mainly through the work of Hewlett and Hibbert (1967), and Dunne and Black (1970), that another mechanism of overland flow generation arises from rain falling on saturated zones of hillslopes from which water is already emerging.

What is the relationship between the flow paths and the geochemistry of hillslope runoff?

For many predictive purposes in the past, it was considered adequate to use simple empirical models to estimate merely the precipitation excess as the water available for delivery to the channel network. Little attention was paid to investigating spatial variability in runoff production in river basins or issues such as the spatial distribution of delivery points to the channels. More recently, however, questions have arisen concerning the acidification of streams and the contamination of ground and surface waters by pesticides, herbicides, and fertilizers, which have underscored the need for the investigation of the space-time structure of river runoff to the channel network as a result of rainfall and snowmelt. This need has intensified debates about each process of runoff generation at the smallest contributing catchments, such as the role of flow in macropores, the effect of microtopography and vegetation on surface flow hydraulics, and the relative importance of random and trending spatial patterns in land surface and near-surface characteristics. Indeed, there has been a continuous advancement in the understanding of the physical processes that control the response of a given hillslope to a precipitation input under certain simplifying assumptions of homogeneity in rainfall, soil, and vegetation characteristics. Further advances are expected from catchment-scale field studies such

IRRIGATION AND DRAINAGE—SUSTAINABLE FOOD PRODUCTION AT WHAT PRICE TO SOCIETY?

Irrigated agriculture has produced food for civilizations that have come and gone during the past thousands of years. Fertile valleys between the Tigris and the Euphrates rivers, the Nile Delta, the Indus plain, the Huanguaihei basin, and the Imperial and the San Joaquin valleys of California are excellent examples of areas where irrigated agriculture has thrived—and then failed. Today, roughly 25 percent of the value of all crops in the United States comes from irrigation on 10 percent of the land farmed. Over 24 million hectares of irrigated land produce more than 50 percent of our vegetables and the majority of our fresh fruits and nuts. In fact, irrigated regions of the western United States produce 40 percent of the table food consumed in the country as well as much that is exported to other countries. There is no doubt about the benefits of irrigated agriculture: abundant yields, short-term security against droughts, and diverse opportunities for farm management that bring a return on private investment with benefits to the producer, processor, distributor, and consumer.

On the other hand, each irrigated acre often uses 1 million gallons of water a year—water that might be used in other ways for the benefit of society. Moreover, irrigation degrades water quality, even with proper management. Irrigation over the course of decades to centuries has caused the accumulation of salts in soils to a level harming crop production. About 10 percent of the total surface of arable lands in over 100 countries of the world suffers from soil salinity and from poor or inadequate opportunity for proper drainage. Without proper management, soils become water-logged, salinized, and nonproductive. Even with good management, including natural or constructed drainage systems, the drainage water carries unwanted salts and agrochemicals and their metabolites, which require disposal. Where are these disposal sites for the drainage waters produced by irrigated agriculture? The oceans, inland seas, deep underground storage, and evaporation ponds (even with dilution by high-quality waters) have all been suggested as alternatives—but our society has yet to decide how we will deal with irrigation and its side effects over the long term.

It cannot be taken for granted that the use of land and water for irrigated agriculture is the preferred use. Permanent irrigation for the production of food and fiber demands two sacrifices. First, the storage and diversion of water for agriculture reduce the water available for other uses. Diversion of water not only modifies the original ecosystem, but also affects the recreational and economic value of the water. Second, the waste product of irrigated agriculture—degraded drainage water containing unwanted pollutants and salinity—must be conditionally accepted and the costs borne equitably.

IRA S. BOWEN
(1898-1973)

One of the better and more robust methods for determining evaporation is based on energy budget considerations at the earth's surface. A major obstacle to its application was, for a long time, the uncertain nature of the turbulent flux of sensible heat in this budget. The matter was finally formulated correctly by Ira S. Bowen.

Bowen was born in Seneca Falls, New York, in 1898. After graduating in physics from Oberlin College in 1919, he went to the University of Chicago to commence graduate studies with Robert A. Millikan, who won a Nobel prize for his determination of the charge of the electron. In 1921, Bowen followed his mentor to the California Institute of Technology, where he was appointed instructor in physics. In 1926, he obtained a Ph.D. in physics at that institution and continued to teach there until 1945. He subsequently became director of two of the world's largest observatories, the Mount Wilson Observatory (1946-1964) and the Mount Palomar Observatory (1948-1964). He died in 1973.

Bowen's main contributions were in astrophysics. In 1927, he presented perhaps his most important discovery, which made him instantly a world expert on gaseous nebulae. On the basis of spectroscopy, he produced what many consider to be the final piece of evidence needed to show that there are no elements elsewhere in the universe that are unknown on the earth.

Bowen's foray into hydrology was brief but epoch making. Turbulence theory was still in its infancy, but Bowen saw correctly that since the coefficients for water vapor diffusion and for heat conduction are nearly the same in air, the convective transport of these admixtures would also be similar. Although Wilhelm Schmidt had postulated this similarity some 10 years earlier, he had not made use of it in his own applications, and it was not readily useful in the energy budget method. Bowen's insight allowed him to express the ratio of the turbulent transport of sensible heat in the air and that of water vapor, in terms of easily measurable variables. The ratio of these two flux terms, or "Bowen's ratio," as it is now called, has been one of the key elements in the successful application of energy budget techniques to determine evaporation of natural surfaces. It is widely applied today, and it has also led to further developments in the field.

as studies of the use of chemical and isotopic tracers (especially ^{18}O and ^{15}O) to trace flow paths of water delivery to channels. Also required are models of runoff production that use information that can be measured and incorporated from field studies.

In river basins, at spatial scales larger than the smallest contributing catchments, significantly new issues emerge regarding runoff production. Basins contain a very large number of hillslopes, and these hillslopes typically display a large overall spatial variability in precipitation, topography, soil, and vegetation properties. Therefore knowledge of the individual process of runoff production from a single hillslope—which is both necessary and valuable—by itself cannot be expected to lead to an understanding and prediction of how river runoff is organized in basins over a wide range of space and time scales. For instance, at the larger scales the space-time variability in precipitation becomes an extremely important variable in governing runoff production. Consequently, statistical methods become crucial in measuring and interpreting field observations and in modeling.

What are the scaling properties of river runoff in basins, and how are the space and time scales connected?

Recent analysis of data on spatial variability in river runoff in basins suggests the presence of multiscaling properties in runoff (see the last section of this chapter). Therefore an important problem is to identify and test the precise nature of scaling in averaged spatial runoff over successively larger scales. Physical connections between scaling models of river runoff and other physical variables such as rainfall, soil properties, vegetation, and topography constitute another important area of research in understanding rainfall-runoff relationships. Long-term research on this topic includes investigations regarding connections between space and time scales governing river runoff.

River Basin Evaporation

Precipitation is the ultimate source for replenishment of the earth's waters. In turn, all the atmospheric water required for precipitation is supplied by evaporation. Evaporation is a major component of the hydrologic cycle, as on a global basis its temporal average is equal to that of precipitation. Even on the land surfaces, it still amounts on average to about 60 percent of precipitation. Moreover, because of the large latent heat involved in the vaporization of water, evaporation allows the transfer of large amounts of energy. For this reason, it has a great impact on the human environment and climate. The partition of the available energy at the earth's surface into evaporation and sensible heat flux is dictated by the nature and the state of the vegetation and the attendant soil moisture stresses.

In spite of the importance of the problem, in current hydrologic practice there are no operational methods available to measure, forecast, or otherwise determine the actual evaporation from river basins. Available representations of physical surface processes, related to evaporation in models, are crude and of limited applicability. Thus it is difficult, if not impossible, to determine the relevant parameters a priori.

One of the major milestones in the development of our present understanding of evaporation was the work done by John Dalton in Manchester, England, in the early nineteenth century. He proposed that the rate of vaporization of a wet surface is proportional to the vapor pressure deficit between the air and the surface, and that this proportionality is a function of wind speed. The eddy-correlation method for measuring evaporation follows the early work by Osborne Reynolds, again in Manchester, England, in the 1870s. The energy budget method was pioneered by Homen in Finland toward the end of the last century and further developed through contributions by Bowen in the United States in the 1920s and Albrecht in Germany in the 1930s. The profile method, used to measure evaporation, has been developed mainly in response to advances in similarity formulations for turbulent boundary layers in the 1930s by Prandtl and von Karman and later in the 1950s by Monin, Obukhov, and others. Progress in evaporation theory has followed developments in radiative heat transfer and fluid mechanics and transport phenomena in turbulent flow. Much of the present activity related to evaporation and transpiration from surfaces covered with vegetation can be traced back to the seminal contributions by Penman and his co-workers in England in the 1950s (for further discussion of evaporation theory and applications, see Brutsaert (1982)).

How do we parameterize (or represent in models) the effects of vegetation and its biophysical mechanisms on the evapotranspiration from a river basin?

Major areas of the earth are covered by vegetation, in varying states of development, senescence, and water availability. Unlike the

HOWARD LATIMER PENMAN
(1909-1984)

A scientist can leave his or her mark on a field in one of two ways: through a startling new discovery or theory or through long-term diligent contributions. Howard Penman was of the latter school, a man renowned for his lifelong work on evaporation rates.

Penman was a distinguished leader in two fields—agricultural meteorology and hydrology—but his career began in pure physics. In his first research position at the Physics Department at Rothamsted Experimental Station in England, Penman was simply told, in remarkable contrast to today's programs, to find something interesting to do. He soon became involved in estimating annual evaporation from bare soil from the difference between rainfall and drainage. His first interest proved to be enduring; except for a period during World War II, Penman stayed devoted to the study of the physics of evaporation from leaves, crops, forests, and catchments for his entire career. His work was important worldwide in the planning of irrigation schemes, in catchment hydrology, and in the whole field now known as land surface processes.

Penman's efforts provided important tools for agricultural meteorology. By eliminating surface temperature from equations that described heat and vapor transfer from open water, bare soil, or vegetation, Penman provided an expression for the rate of evaporation from wet surfaces as a function of standard weather variables. His careful analysis of evaporation measurements helped develop values for several essential constraints, and his 1948 paper describing this endeavor, "Natural Evaporation from Open Water, Bare Soil, and Grass," is a classic.

Penman became chairman of the Physics Department at Rothamsted in 1954 and held the position until his retirement 20 years later. His professional honors were many—in 1952, he was awarded a Darton prize by the Royal Meteorological Society, and in 1966 he received the Hugh Robert Mill medal for his work on the water balance of the earth's surface and its application to agriculture, among other awards. But apart from his dedication to his field and his contribution of the Penman formula, he is remembered for his ability to inspire younger colleagues and for his humor. He was a genial and gentle man, whose criticism of a colleague's paper would often begin, "Sir, may I ask a question from a position of complete ignorance?" only to be followed by a probing, perceptive analysis.

underlying soil, vegetation is not a passive substrate, but an active transmitter and metabolizer of water. The hydrologic aspects of the concomitant biophysical processes are far from understood. Plant physiologists have been able to elucidate the cellular mechanisms of stomatal control of water vapor exchange between leaves and ambient air. However, this has not yet been translated into effective parameterization to describe the plant canopy in the hydrologic context. This is not surprising: descriptions at a given scale cannot be obtained by mere superposition or addition of known phenomena at the scales one or several steps below. Indeed, an entirely different conceptualization may be needed.

A useful concept being applied for this purpose is the bulk stomatal or canopy resistance. This parameterization is usually intended to express the control on transpiration exerted by the plant for certain conditions of moisture availability. Numerous experiments have been conducted to determine the stomatal or plant resistance for various types of vegetation and climate. Although the concept is effective as a diagnostic index, it is still very difficult to use for prognostic purposes in hydrology. Understanding the intervention of active vegetation, within the partly saturated soil-turbulent atmosphere complex, is one of the major pressing problems in this field.

How do surface radiation balance and boundary layer dynamics control land surface evaporation at the mesoscale?

The region of the atmosphere that is most directly affected by the land surface is referred to as the atmospheric boundary layer. It extends up to elevations of about 1 km, and within it occur all the physical processes that control the exchange of water, energy, and momentum between the atmosphere and the surface. Predominant among these processes are the surface radiation balance and the atmospheric turbulent transport. As far as the latter is concerned, an important property of this layer is that the characteristic vertical length scales are typically 2 orders of magnitude smaller than the horizontal ones. Thus atmospheric observations made a few meters above the surface can be representative only of fetches on the order of hundreds of meters; this represents roughly the so-called field scale. In a similar vein, it is clear that measurements taken higher up in the boundary layer may be expected to be representative for surface conditions over larger areas extending upwind over distances of the order of tens of kilometers; these characterize the meso-gamma scales (2.5 to

25 km), which are typical scales for source watersheds of similar hydrologic regions. The much-needed progress in the formulation of parameterizations for regional evaporation and related surface fluxes will likely depend on a better understanding of both boundary layer dynamics and surface radiative properties. Developments in this field will be significant for hydrology.

Currently, natural evaporation is not being measured operationally. Yet it would be useful to set up a worldwide network of land-based index stations where surface energy fluxes and turbulence parameters were measured systematically and routinely. As an initial goal, the density of this network could be designed to be of the same order as that for the upper air (radiosonde) observations gathered for synoptic purposes. The anticipated deployment of various sounders and profilers for the lower atmosphere, as planned for the coming decade by the U.S. National Oceanic and Atmospheric Administration, may provide additional opportunities to accomplish this. Remotely sensed data from satellites have great potential in this regard and should be developed.

In water resources planning and management, catchment evaporation must be known since it is a consumptive use and cannot be recovered. During periods of drought, soil moisture is one of the main measures of water availability, and evaporation is one of the main depletion mechanisms. However, for many flood situations there is strong evidence that an important factor governing flood severity is the infiltration capacity of the catchment; the capacity to store water in the soil profile depends largely on the soil water content and hence on the antecedent evaporation from the watershed as a whole.

During the past 20 years, continuous progress has been made in the development of physical-dynamical models to describe the general circulation of the atmosphere as it affects the evolution of weather and climate. The theoretical basis for these models, discussed earlier in this chapter, consists of the relevant thermohydrodynamical equations for the atmosphere (and the oceans) and suitable formulations for the boundary conditions, including those to calculate momentum, heat, and water vapor fluxes. One component of these GCMs, to which they appear to be quite sensitive, is the proper formulation of the hydrologic budget of the land surfaces of the earth, namely, soil moisture, evaporation, and related variables. For example, numerical experiments have shown that large-scale changes of land surface evaporation in such models produce significant changes in the predicted atmospheric circulation and precipitation. This and other evidence indicates that there is a critical need for sound parametric expressions for evaporation and related land surface processes over areas with scales of 10 to 100 km.

Surface and Subsurface Water in a Freezing Environment

How do heat and mass flow in frozen soils control water seepage in such media?

What mechanisms characterize the effects of pollutants in the hydrologic behavior of freezing soils?

A large fraction of the earth's land surface freezes seasonally, and about one-fourth of the land contains perennially frozen ground (permafrost) that partially thaws in summer. The hydrology of these cold regions is dominated by freezing and thawing processes and by the presence of ice on and below the surface. Surface water and ground water interactions involve the usual multiphase porous media flow, coupled with heat flow, phase changes, and other complications such as solute rejection, water pumping, and ice segregation. This complexity challenges our understanding, but research is needed because of the impacts of these problems on hydrologic and water resources applications, geotechnical considerations, the development of cold-region landscapes, ecosystem dynamics, and biogeochemical cycling.

The unusual nature of cold-region landscapes was noted by early explorers: polygonal networks of ice wedges, mountain-like pingos, moving soil aprons, vast fields of icings (*Aufeis*), and similar features attracted the attention of geologists and geomorphologists. By the beginning of this century, it was realized that the peculiar motion of water in freezing soils played a vital role in producing these structures, but quantitative experimentation and theoretical investigations did not become important until World War II and thereafter, when the interests of defense activity and petroleum exploration brought engineers to the high latitudes and the Soviet experience with permafrost became known in the West.

Driven largely by geotechnical considerations, engineers built up a satisfactory general understanding of the processes involved in soil freezing, but predictive ability was limited. Thermodynamicists, soil physicists and chemists, and quantitative geomorphologists, as well as hydrologists, have become interested in the processes, and now a basic understanding of many of these phenomena is developing. However, owing mainly to the complex physics involved in the coupling of surface and ground water systems, the field is still underdeveloped scientifically.

Infiltration into frozen or partly frozen soil involves complicated interactions. The hydraulic conductivity, for instance, depends not only on the physical properties of the soil and the fluid, but also on variables such as temperature and the amount of ice in pore spaces. Ice lenses restrict infiltration during spring melt, influencing surface water runoff and ground water recharge. These ice lenses form behind the freezing front, separated from it by a "frozen fringe" through which water is transported from the unsaturated, unfrozen soil below. The hydraulic conductivity in the frozen fringe is one important factor in determining where and how much ice segregation will occur. The capillary pressure gradient depends on the freezing process and may be an order of magnitude greater than the gravitational pressure gradient. Because of this, the net flow of water in the soil is generally from warmer to cooler areas, in the direction of the heat flow. Solutes affect the freezing point, and on freezing, solutes are ejected and concentrated in the soil water. The structure of the soil is also affected by freezing and ice segregation and by thaw compaction, causing subsequent changes in the hydraulic conductivity.

An extreme case may result where frozen ground persists for years to millennia. Massive horizontal ice beds, from one to many meters thick, may form at the base of fine-grained soils underlain by sand. In addition, repeated expansion and contraction of the near-surface materials caused by temperature fluctuations may produce vertical ice wedges, resulting in a complex array of permeable and impermeable structures.

Water in the permeable, unfrozen layers may be trapped under high pressure, so that any disruption of the confining system can lead to spectacular results. For instance, liquid water may intrude, like a laccolith, into the soil, resulting in the growth of a conical body of ice under the turf mat (a pingo). Or water may be trapped in unfrozen soil between frozen ground below and a freezing front moving down from the surface; this can occur in stream channels or on slopes. The pressurized water may break through the confining layer and pour out onto the surface, forming icings that may extend for tens to hundreds of kilometers. Although a basic understanding exists of why these phenomena occur, the ability to predict where, when, and to what extent they will occur is limited by our lack of quantitative understanding of heat and mass flow in freezing soils.

A related problem is the motion of water through cold snow. This influences the timing of runoff, extent of soil moisture, and rates of ground water recharge, as well as the concentration of atmospheric pollutants ("acid snow") delivered to the earth's surface during spring melt. Improved understanding of the snow problem could aid in the solution of the soil problem, as similar processes of heat and mass transfer are involved in both.

Many opportunities exist for scientific advancement in this field of surface water and ground water interactions in the presence of freezing. Perhaps the most critical and challenging aspect has to do with the microphysics of the transfer of water through soil to and into a freezing front and with the segregation of this water into ice lenses. Related problems include the physical and chemical processes involved in solute motion and the effect of overburden pressure on the state of water (as ice and liquid) and the soil matrix. These basic processes will have to be addressed on all fronts: by theory, laboratory experimentation, in situ field experiments, and numerical modeling. Special attention needs to be paid to spatial variability and the problem of scaling. The results of a better understanding can then be used to parameterize larger-scale models and thus obtain solutions to practical problems.

Basic hydrologic data for the cold regions of the world are sparse. Even such fundamental information as streamflow and ground water levels is rare, partly because there has not been an active user constituency demanding such information, but largely because it is difficult to collect such information in this harsh environment. A vast spectrum of needs in the high latitudes would be served if data collection in the following areas were pursued:

1. Hydraulic conductivity in different soil materials, as a function of temperature, pore ice content, and other conditions; relation of water saturation to capillary pressure, for different soils and ice contents; and changes in solute content and movement in different soils for different freezing scenarios and different chemical species.

2. Pressures before, during, and after freezing in liquid water, ice lenses, and the soil matrix.

3. Liquid water velocities and fluxes through different frozen soils and snow with differing driving potentials.

4. Basin studies, perhaps in coordination with interdisciplinary research areas such as biosphere observatories, to determine energy and mass in different phases of the hydrologic cycle.

5. Remote sensing, to scale up observations of soil moisture (if possible) and other hydrologic variables from detailed study sites to basins and even regions.

Although understanding of the interactions between surface and subsurface water in freezing conditions is needed for improved knowledge of our earth system, it would also benefit a broad spectrum of practical applications:

• *Biogeochemical cycling* in cold regions depends on the ground water table and other hydrologic conditions in summer, and these depend in turn on freezing and thawing processes. Whether high-latitude peatlands and wetlands are sources or sinks of carbon dioxide and methane depends on the aerobic versus anaerobic microbial processes, which depend on water table level. The flow of dissolved organic carbon and nutrients within and between high-latitude terrestrial ecosystems and from them to marine ecosystems also depends on the processes of freeze and thaw.

• *Global climate* depends on the surface radiation balance as well as evaporation, transpiration, soil moisture, and other factors. The surface radiation balance in high altitudes is characterized by the presence of highly reflective snow, and the persistence of snow cover involves interaction (heat and mass flow) with the substrate. Soil moisture, evaporation, and transpiration depend on infiltration and drainage, which are controlled in cold regions by freezing, ice segregation, and thawing processes. One of the best records of climate change in the past decades to past century is found in the temperature distribution in perennially frozen ground, but full use of this information involves better understanding of heat and mass flow through freezing soils.

• *Studies of the global hydrologic cycle* in a greenhouse-affected climate will require knowledge of potential long-term changes in some of the major global reservoirs. Land ice masses (glaciers and ice sheets) will most likely diminish in volume, causing sea level to rise. Any estimate of the timing of this effect requires an understanding of the flow of water in cold snow, which is related to the problem of understanding the flow of water in cold soils.

• *Water resources development* in arctic and alpine lands will be aided if many problems caused by freezing can be solved. Water quality is affected by solute rejection during freezing, ground water between or under perennially frozen layers may be under high pressures, and modification of these systems by humans may lead to untoward results unless a proper understanding of heat and mass flow in freezing soils is at hand.

• *Geotechnical problems* such as the stability of pilings, structures, and transportation arteries continue to plague development in cold regions. The main problem is frost heave, caused by the buildup of ice lenses and other ice segregations in frozen soils. Obviously, a better understanding of heat and water flow in these frozen soils and in the frozen fringe is critical to the improvement of geotechnical engineering.

Hydrology of Snow-Covered Areas

The distribution of snow deposition is not uniform, and the processes of snow metamorphism and melt proceed at different rates in different parts of a single drainage basin. Seasonally snow-covered areas of the earth, especially in mountain ranges, are important components of the global hydrologic cycle, even though they do not cover a large portion of the earth's surface area. They are the major source of water for runoff and ground water recharge over wide areas of the mid-latitudes, they are sensitive indicators of climatic change, and the release of ions from the snowpack is an important component in the biogeochemistry of alpine areas. Alpine watersheds are particularly sensitive to damage from acidic deposition because they are usually weakly buffered and their acid-neutralizing capacities are limited.

Examination of the chemical and nutrient balances of such watersheds is difficult, however, because their hydrologic characteristics are only partially understood and difficult to measure. The amount of snow accumulation varies because the rugged topography causes snow to accumulate at varying depths at different elevations and different exposures, and wind and avalanches subsequently redistribute the snow. Once the snow is on the ground, significant variation in the surface climate results from local topographic effects. The major contributor to this variation is solar radiation, although there are also important topographic variations in long-wave radiation, wind speed, temperature, humidity, and moisture in the underlying soil. As a result of these variations, energy exchange at the snow surface proceeds at widely different rates within even small drainage basins, and therefore the rates of the processes of snow metamorphism and melt also vary. Part of the basin may be melting, releasing water and soluble ions to the soil, whereas another, shaded part of the basin may still require significant energy input to bring the snowpack to 0°C. Our knowledge of snow processes at a single field site where measurements are available must be integrated over the drainage basin.

This integration over larger areas is difficult because data collected at a meteorological station or snow courses will seldom represent conditions throughout an entire drainage basin. Intensive field sampling of snow properties is possible only in very small areas and even then only for research purposes. The only method of obtaining widespread measurements is through remote sensing of the snow cover, from aircraft or satellite. Therefore we need to understand the relation-ship between the physical characteristics of the snowpack and its electromagnetic properties.

How can we integrate the radiation balance over large areas to provide estimates of times and rates of snowmelt in alpine terrain?

In a seasonal snow cover, newly fallen snow is thermodynamically unstable, undergoing continuous metamorphism until melt occurs in the spring. These metamorphic changes and melting, along with chemical fractionation, are driven by temperature and vapor gradients within the snowpack and by energy exchange at the snow surface and at the ground. In the absence of strong temperature gradients in the snowpack, dry snow acts to minimize its surface free energy, and slow grain and neck growth occurs by local vapor diffusion and heat flow. On the other hand, strong temperature gradients imposed on the snow cover cause spectacular changes in grain morphology and grain number density, depending on the temperature, the magnitude of the imposed gradient, and the initial density. As the snow recrystallizes, the density often remains constant. Some crystals grow at the expense of others, causing a decrease in the number of crystals per unit volume. The crystal forms resulting from snow metamorphism have widely varying shapes, sizes, and continuity, which affect mechanical, radiative, and chemical processes in ways that are poorly understood.

By accounting for energy fluxes to the snowpack, one can estimate the temperature profile of the pack and account for loss of snow mass through sublimation and melting. The spatial and temporal distribution of these processes in the basin could intensify or attenu-ate any ionic pulse in the meltwater. Loss of snow directly affects the solute concentration of the remaining snow, and the spatial and temporal distribution of the onset of melt within a watershed further determines the effects of the ionic pulse. The distribution of chemical species with depth in the snowpack—that is, the initial conditions for melt—has less influence on the chemical hydrograph than does the grain-scale distribution of solutes. Knowledge of energy fluxes to the snowpack, combined with bulk snow chemistry measurements, should therefore provide a method for estimating the chemistry of runoff to aquatic systems.

Most energy exchange studies, in a variety of snow cover conditions and locations, have shown that usually the radiation balance is the dominant term. Recent research in solar radiation absorption by snow has involved attempts to account for terrain effects in mountainous areas and to measure the spatial distribution of albedo. The hope is that the radiation balance can be integrated over a drainage basin to provide more accurate estimates of the time of initiation of snowmelt and subsequent rates than are possible with degree-day methods, which are accurate enough in forested areas but often unreliable in alpine terrain.

The long-wave portion of the spectrum—the radiation emitted by the atmosphere and surrounding terrain—has received less attention. In alpine regions the places most likely to have a positive radiative energy balance early in the snowmelt season are near the valley bottoms because radiation is emitted from the surrounding valley walls throughout the day and night. In forested environments the long-wave irradiance accounts for the major portion of the snow surface energy exchange; hence the good correlation between snowmelt and degree-day variables in such areas.

The radiation balance usually accounts for most of the energy exchange for snowmelt and is the source of most of the spatial variation, but sensible and latent fluxes can be large. Usually they are of opposite sign; under most conditions, sensible heat exchange adds energy to the snowpack and latent heat exchange removes it. However, this is not always the case, and the topographic distribution of the contributing wind speed is a significant unsolved problem.

What is the three-dimensional distribution of the concentration of chemical species and particulate and colloidal contaminants in the winter snowpack, and how does it change throughout the season?

The amount and chemical composition of water entering streams and lakes in alpine areas depend on the quantity and composition of atmospheric deposition of water and chemical species and on the hydrologic and biogeochemical processes occurring in the snowpack and in the drainage basin. Understanding these processes requires a coordinated, interdisciplinary effort.

First, we must understand the spatial distribution of the rates of snowmelt and metamorphism. This will require innovation both in measurement techniques and in modeling of the processes.

Second, we must be able to estimate the spatial and temporal distribution of rates of chemical elution throughout a watershed. This will require data on the initial distribution of chemical species and knowledge of their elution from the snowpack in response to melt and metamorphism.

Third, biogeochemical processes in the watershed must be integrated with our knowledge of the snow hydrology. The processes in the snow interact with those in streams, lakes, sediments, soils, and vegetation.

How can a combination of remote sensing data and field surveys yield the most reliable information on the distribution of the snow resource and on the rates of snow metamorphism and melt over alpine watersheds?

In the visible and near-infrared wavelengths, from 0.4 to 1.0 μm, snow is the brightest substance in nature, and its reflectance is rivaled only by that of clouds and bright soils. In satellite imagery, delineation of the snow-covered area is usually straightforward, and clouds can be distinguished from snow if the correct wavelength range is available, from 1.55 to 1.75 μm. Thus the principal use of remote sensing of snow in these wavelengths has been to map the extent of the seasonal snow cover. Throughout the world, in both small and large basins, maps of the snow cover throughout the snow season are used to forecast melt. These maps are useful both in areas with excellent ancillary data and in remote areas with few supporting measurements.

Since the first mapping of snow cover from a satellite, the spectral and spatial resolution of the available sensors has improved significantly. Emissivity in thermal infrared wavelengths is not very sensitive to snow properties, and therefore measurement of snow surface temperatures from the thermal infrared emission of the snowpack requires only correction for atmospheric attenuation of the signal. However, reflectance in the visible wavelengths occurs over the top 10 to 50 cm of snow, depending on the grain size and density, and emission in the thermal wavelengths comes from the top few millimeters. Therefore neither wavelength region can provide measurements of the snow water equivalence.

In fact, only in the microwave frequencies can we estimate snow water equivalence from satellites. Microwave remote sensing can be accomplished by measuring either emitted radiation or the intensity of the return from a radar. Microwave measurements have the capability to penetrate clouds and thereby permit observations of snow under nearly all weather conditions. For instance, the Nimbus series of spacecraft and the Defense Meteorological Satellite Program have provided data on global snow cover since 1979.

The relationship between hydrologically important characteristics of a surface and its electromagnetic signature, reflected or emitted, is one of the central issues in remote sensing for wavelengths from gamma radiation to the microwave. Different properties of snow affect the electromagnetic signal at different parts of the spectrum, but present understanding of these relationships must be sharpened to be fully practical. The electromagnetic properties usually are investigated at the scale of a study plot (a few meters), but the application of the findings is essential to solving problems at much larger scales—from drainage basins (10 to 100 km) to continents (1,000 to 10,000 km).

Current work on the electromagnetic properties of snow in the visible through infrared wavelengths fails to consider two issues. The angular distribution of the reflectance has not been fully investigated, and we do not know the relationship between the grain sizes that we interpret from measurements of the reflectance and the physical properties. Similarly, in the microwave frequencies, the major difficulty in experimentally testing models of emission and backscatter is that the snow property observations made by field scientists are usually inadequate to determine the theoretical model parameters to allow comparison with radiation measurements. In the mountainous regions especially, our interpretation of the properties of the snowpack from measurements of the electromagnetic signature is not precise. Therefore comprehensive measurements of radiometric properties should be combined with precise measurements of snow properties.

Unfortunately, we do not know how to measure some of the important snow properties. For dry snow, stereological methods can be used, whereby the snow sample is saturated with a supercooled fluid, which is then frozen. Such samples can be sectioned and photographed, from which dimensions and shapes of grains can be obtained. In wet snow, however, theoretical calculations show that the electromagnetic behavior of snow in the microwave frequencies is sensitive to the geometry and volume fraction of the water inclusions, but we have no method for measuring the shapes of the water inclusions. Thus future progress in the use of remote sensing to measure snow properties of hydrologic interest will require scientists who are well versed in radiative transfer theory, so that they can analyze the relationship between the physical and the electromagnetic properties of snow.

Evaporation from Large Water Bodies

Oceans, lakes, reservoirs, and other large inland water bodies constitute a valuable resource available to society. More than 70 percent of the earth's surface is covered with water. To preserve these natural resources and to allow their optimal use, a better understanding of the physical properties of such large water bodies and of their interaction with the environment is required. Of critical importance is the mass and heat exchange at the interface between the water and the surrounding atmosphere.

Most methods available are based on similarity formulations for the profiles of mean wind speed, temperature, and humidity above the water surface. These formulations are usually in the form of bulk transfer equations, some with adjustments for atmospheric stability and for waviness of the water. These methods can give good results when the water surface temperature is available. Methods based on energy budget considerations are less commonly applied. Over the ocean the necessary measurements can be obtained from instruments aboard research vessels or mounted on buoys. In the case of lakes and reservoirs the data often are measured onshore, but these require some type of calibration or transformation to derive mean surface fluxes for the entire water surface.

The main problem in ascertaining evaporation from the oceans of the world is the general dearth of data. For lakes and reservoirs, advection effects caused by the proximity of the surrounding landscape create serious difficulties.

Evaporation from the oceans and from the lakes has fascinated human beings since prehistoric days. Best known perhaps is the description in Ecclesiastes 1:7, from the fourth century B.C.: "All streams run into the sea, yet the sea never overflows; back to the place from which the streams ran they return to run again." This seems to have been a universal preoccupation, because similar descriptions have appeared in writing throughout the ages, starting with ancient Chinese, Indian, and Greek texts. In the seventeenth century, Edmund Halley, geophysicist *avant-la-lettre* of comet fame, was concerned with evaporation of the sea and obtained an estimate by experiment. But again, it was John Dalton's work at Manchester, England, at the beginning of the nineteenth century that provided the initial impetus for more fundamental approaches. More recent advances in the development of heat and mass transfer formulations generally have followed the formulations in turbulence similarity by such well-known scientists in fluid mechanics as Prandtl, von Karman, Taylor, Obukhov, and Monin. Sverdrup in the late 1930s was one of the first to apply these ideas to ocean evaporation (Sverdrup, 1937, 1946). The difficult problem of advection in lake evaporation was first dealt with in a fundamental way by Sutton in the 1930s (Sutton, 1953). The experimental research of Harbeck at Lake Hefner in Oklahoma in the 1950s has set the standard for much of the subsequent work in lake evaporation (Harbeck, 1962).

What are the microphysical mechanisms of evaporation from a wave-disturbed water surface?

The oceans and other large water bodies are a major source of the water vapor in the hydrologic cycle that falls to the surface of the earth as precipitation. Yet the microphysical mechanisms of evaporation and related transport phenomena from a water surface disturbed by waves are poorly understood. The fluxes of water vapor, sensible heat, momentum, and other admixtures near the water surface are primarily the result of the turbulence in the air, but this turbulent flow interacts strongly with the irregular wave field of the underlying water surface. The turbulence in the air is conditioned by the presence of the moving waves and thus is naturally different from the classical case of flow past a solid rough wall. At the same time, the sea state and the waves, and the distribution of their shapes and sizes (i.e., their spectral characteristics), are normally a direct consequence of the nature and the intensity of the wind and the turbulence in the air. These interactions with strong built-in nonlinearities and feedback mechanisms are usually accompanied by wave breaking. In open seas and along shorelines, this wave breaking is a dramatic surface phenomenon—a force that has long held people's imagination and elicited powerful expressions in poetry, painting, and music.

Wave breaking is responsible not only for the transfer of mechanical energy from the atmosphere to ocean currents and turbulence, but also for the enhancement of the exchange of water vapor (and other gases as well) between the atmosphere and the ocean. Wave breaking, which is manifested by whitecap formation, directly entrains air into the water in the form of bubble plumes. A phenomenon closely related to this is the production of water spray and marine aerosols (salt) as a result of the shearing of the wave tops by the wind and of the bursting of the bubbles in the whitecaps.

To monitor ocean evaporation, it would be essential to deploy a network of buoys, with the necessary instruments to measure the standard variables for the application of profile and bulk mass transfer methods. With the anticipated improvement of satellite monitoring for wider coverage, this buoy network would provide the needed ground truth for anchoring and calibrating the satellite data.

With increasing pressures on the world's water resources, it will become critical to monitor the water budgets of freshwater lakes and reservoirs. Because evaporation is such an important component in these budgets, this will require the installation of proper instrumentation.

Solving these problems will require scientists with general backgrounds in physical science and engineering, and also with a special interest in hydrology, oceanography, and atmospheric science. Water entering the evaporation process from lakes and reservoirs becomes nonrecoverable locally. Thus evaporation is a "consumptive" use, and this fact is fundamental to water resources planning and management. The vapor transfer at water surfaces must be predicted as accurately as possible, because such information is indispensable in designing the capacity of storage and flood-control reservoirs and assessing the value of natural water bodies for municipal and industrial water supply, navigation, irrigation, and recreation.

Large reservoirs, lakes, and the continental shelf have become attractive as sites for construction of power-generating plants and related industries because of the abundance of condenser cooling water. The increase in the past few decades of steam electric power plants makes the need for cooling water especially critical. The environmental impacts of these plants can be evaluated only if the evaporation and cooling rates are known for the ambient waters subject to warmwater effluents.

On a global scale, the oceans with their huge inertia provide the flywheel for the earth's climate engine. Since evaporation is one of the key components of the oceanic energy budget, an understanding of it is essential in any assessment of climate or of climatic change. But oceanic evaporation also plays a critical role in air-water interaction at more local scales. For instance, there is evidence that in hurricane genesis the rate of sea-air transfer is the principal limiting factor governing the ultimate intensity. A satisfactory explanation for this would require more strongly nonlinear transfer coefficients than were hitherto surmised.

HYDROLOGY AND LIVING COMMUNITIES

Introduction

The dependence of life on water is fundamental, for it is a major constituent in essentially all organisms. **The life cycles of most terrestrial and freshwater organisms are organized around their access to water**; seasonal thermal cycles and concomitant chemical changes in the water exert strong influences. The hydrologic cycle—the distribution of water and associated nutrients on the planet in space and over time—and the annual thermal and day-length changes are elements of fundamental physical templates for biological processes.

Throughout history, many of the efforts to understand and predict hydrologic processes, and attempts to manage them, have been motivated by biological concerns—most often our human requirements. An understanding of the hydrologic cycle can provide a framework for interpreting key biological processes. Hydrologic and biological processes interact over the range of spatial scales, from the microscale of small habitat units, through areas the size of major drainage basins (mesoscale), to the macroscale of continents; temporal scales can vary from minutes to centuries. The influences of hydrologic patterns and events on biological processes are pervasive. In turn, **through feedback loops the biological processes may modify the hydrologic setting**. For example, the distribution of vegetation and its productive potential can be directly or indirectly related to climate gradients associated with the amount and seasonal distribution of precipitation. In turn, vegetation affects both precipitation and temperature through transpiration, and forests play a strong role in determining surface albedo, particularly in snow-covered terrain.

Another example of how hydrologic processes can drive biological responses is the transport of organic material and sediments in running water environments. Transport and storage of sediments affect the distribution and abundance of aquatic microbes, plants, and animals. Moreover, the nature of organic coatings on stream sediments is an important factor in the quality of runoff, which in turn deeply affects the stream biota and all the living communities in the watershed.

The hydrologic cycle clearly is fundamental to the patterns of adaptation of terrestrial and aquatic living systems. Its importance arises from the multiple roles through which water acts on natural ecosystems: carrier, cooler, substrate, and mechanical force. The fluxes of most key biological variables (e.g., biomass) and the operating chemical processes (e.g., carbon assimilation) are intimately intertwined with hydrologic processes (e.g., evapotranspiration). Thus the hydrologic cycle represents a fundamental physical template for biological processes. This template presents some of the best opportunities to search for general principles that may guide the organization of living communities. It also offers guiding physical principles at different scales that are invaluable in the study of the interactions among the fundamental processes governing the evolution of living communities.

Many of the adaptations of aquatic organisms—from bacteria to fish—are related to patterns of stream runoff. These include adaptations for maintaining position in flowing water or avoiding the thrust of the current. However, the most significant adaptation of freshwater organisms is the matching of their life cycles to variable hydrologic events. That is, biotic populations in freshwater environments have evolved to survive and perpetuate, in the long term, in response to stochastic hydrologic patterns. Thus many biological processes respond to hydrologic events and may even be driven by them. However, as stated in Chapter 2, the committee's definition of hydrologic science is restricted to processes that are significantly interactive with the hydrologic cycle. That is, the biological processes of interest here are not just affected by hydrology but in turn affect hydrology in a significant manner.

Some Frontier Topics

Physiological Explanation of Boundaries of
Major Vegetation Formations

What is the physical basis for the geographical distribution of the major vegetation types on the earth's continents?

The earth's vegetative cover is coupled physically to the climate system and to the soil by the fluxes of thermal energy, moisture, and nutrients. Different vegetation types have different physiological needs and tolerances that develop over time in a process of co-evolution with their soil and climate partners. The current manifestation of this evolution and migration, as modified by human activity, is the distribution of vegetation formations. Figure 3.21 presents an example for eastern North America. What are the physical reasons for a particular type of vegetation being confined to a certain geographical zone? For example, going south at constant longitude, why should the forest type change from evergreen (boreal) to deciduous and then back to evergreen (southern pine) again? It is important to gain quantitative understanding of these boundaries if the effects of climate change are to be forecast accurately.

The transitions from one vegetation type to another are known as ecotones. Because they represent marginal conditions, ecotones are especially sensitive to changing climate, and predicting their location is an important test of the atmospheric general circulation models.

Researchers have recognized the relationship between the distribution of vegetation and climate since at least the beginning of this century. Most of this work is empirical, however, and it correlates the location of vegetation type with climatic parameters such as temperature, precipitation, and potential evapotranspiration without specifically considering the causal physical relationships. Some in-

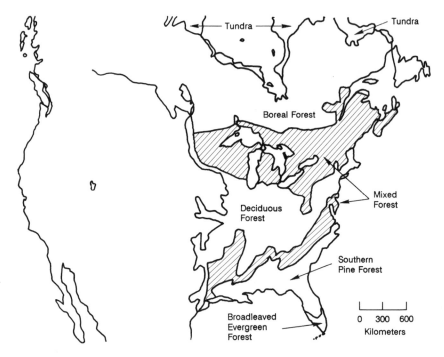

FIGURE 3.21 Forest formations of eastern North America. SOURCE: Map from Little (1971) courtesy of the U.S. Department of Agriculture. Boundaries, by permission, from Eyre (1968); copyright © 1968 by Edward Arnold.

vestigators have used such correlations as the basis for predictive models, but there is no guarantee that they will hold as climatic conditions change.

A more recent modeling direction "grows" individual trees in a stand incorporating both competitive interactions and climatic constraints. While these models have successfully simulated both past and present vegetation distributions, they contain a great many parameters that are difficult to evaluate.

Stepping back from the individual tree and the complexity of its interactions with its neighbors to take a wider view of the forest admits simpler physically based models that can capture the primary distributional features of a given community. For example, modeling savanna vegetation as a tree-grass system where the two vegetation types compete for water and solar energy has provided insight into the conditions for both equilibrium and stability of such communities.

Models of this last, competitive type may help forecast ecotone location even where water is not limiting. However, since they are essentially equilibrium models, their applicability requires that the time scale of climate change be small with respect to that of adaptation and yet large with respect to that of migration. The basic premise is that the relationship between different vegetation types is competitive and that the prevailing environmental conditions are important in determining vegetation distributions only as they affect the relative competitive abilities of the plants. Where water is not limiting, dominance in a given environment may go to the plant type that produces the most biomass and hence wins the race to grow and reproduce. Biomass production is a function of carbon assimilation through the process of photosynthesis. Since both carbon dioxide and water vapor exchange occur through the stomata, the evapotranspiration rate is often used as the basis for estimating productivity (Figure 3.22).

Optimality Constraints on Vegetation Communities

Do adaptive or evolutionary pressures lead plant communities to states of development or to physical characteristics that are in some sense optimal with respect to a critical dimension of their environment?

Ecologists studying the population dynamics of plant communities have long noted the existence of environmental conditions called ecological optima under which certain species occur most abundantly in nature. For the controlled conditions of the laboratory, where competition among species can be prevented, these optima are called physiological optima (Figure 3.23). Observations of such optima lead to speculation about whether there are preferred operating points of the climate-soil-vegetation system toward which it is driven by natural selection and/or adaptation, and at which it is stabilized by inherent (homeostatic) feedback processes.

In 1974, M. I. L'vovich presented the variability of the components of the mean annual water balance as a function of hydrophysical properties of the soil (Figure 3.24). Because of the relationship between evapotranspiration and plant productivity (see Figure 3.22), this takes on the characteristics of an ecological optimum. It has led subsequent investigators to suggest that soil building and plant community development may occur synergistically over time along a path that carries the soil-vegetation system to the point of maximum evapo-

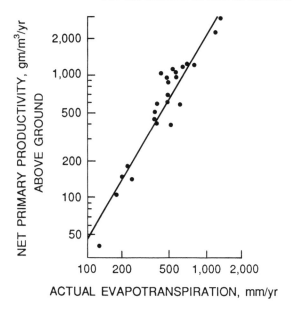

FIGURE 3.22   Net primary productivity in relation to actual evapotranspiration. Reprinted, by permission, from Whittaker (1975) after Rosenzweig (1968). Copyright © 1968 by the University of Chicago.

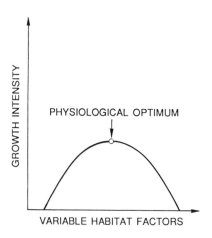

FIGURE 3.23   Idealized ecological optimum. SOURCE: Reprinted, by permission, from Eagleson (1982). [After Walter (1973). Copyright © 1973 by Springer-Verlag, Heidelberg.]

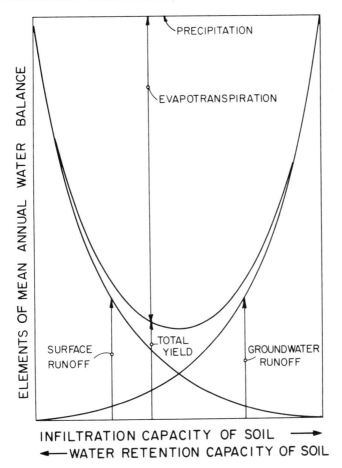

INFILTRATION CAPACITY OF SOIL ⟶
⟵WATER RETENTION CAPACITY OF SOIL

FIGURE 3.24 Variation of water balance with soil characteristics (total yield equals surface runoff plus ground water runoff). SOURCE: Reprinted, by permission, from Eagleson (1982) after L'vovich (1979). Copyright © 1979 by the American Geophysical Union.

transpiration and that this is the so-called climatic climax state of the community.

Where vegetation types having similar physiology are present at the same site, competition for resources will occur. At the geographical boundaries between communities of different vegetal type, such as the boreal-deciduous ecotone discussed earlier, we might expect the states of the two communities to be equal, as sketched in Figure 3.25.

When water is the limiting resource in community development, species and fractional canopy covers result that minimize the stress induced in plants by an atmospheric evaporative demand that exceeds

MICHAEL EVENARI
(1904-1989)

Michael Evenari is best known in the field of plant ecology, especially as a pioneer in plant ecophysiology, but his career illustrates how important and interrelated hydrology is to other fields. Evenari was born in Germany in 1904 as Walter Schwarz. As a university student, he studied botany, zoology, physics, and philosophy, and he moved quickly along established routes toward a position as a university professor. But his brilliant start stalled abruptly on April 1, 1933, the day when German Jews were boycotted. Three weeks later, Walter Schwarz arrived at Haifa, Palestine, and began life anew as Michael Evenari.

Evenari became a lecturer at the Jerusalem University but soon was fighting in Italy with the Jewish Brigade. Back home in 1945, he worked with colleagues to establish the Botany Department at the Hebrew University as one of the world's most famous centers for ecology. For more than 50 years, he developed his ideas in ecology, seeking to explain plant establishment, performance, survival, and distribution.

Because of this work, Evenari became intrigued with the challenge of desert farming, and there he made his most important contributions to hydrology. His studies of historical runoff farming under extreme desert conditions were an outstanding and highly original contribution. Through extensive historical research, arduous field work, and aerial surveys, he concluded that a flourishing empire once existed based on runoff directed from extended mountain catchment areas onto fields. He worked to reconstruct such a farm, Avdat experimental farm, and today its green fields are an oasis in the yellow landscape of the Negev Desert, and a unique opportunity for researchers from around the world. Without irrigation, runoff farming proved capable of supporting pastures, vegetables, and fruit trees. Although we are still far from a full understanding of the complicated interrelationship within desert ecosystems, his work was a significant step toward beginning to understand the integration of the desert ecosystem's biological, geological, meteorological, and hydrological features.

Michael Evenari received many honors during his long, productive life—both for his scientific work and for his humanitarian contributions. He was a thoughtful man with a deep interest in the fate of humankind. On receiving the prestigious International Balzan Prize in 1988, which honors deserving cultural and humanitarian work, Evenari gave an acceptance speech that captured some of his special insights. He explained that while we see the ever-increasing advancement of science and technology as progress, we must not forget that the blessings of scientific discovery are not without a price. There is scarcely an invention that man has not turned into a lethal weapon or an instrument of ecological destruction. The ethical behavior of mankind has not kept pace with technical development. Mankind has become its own worst enemy. But Evenari implored us to change our way of life, to reform our educational value system, and to make it our primary goal to educate and practice the ethical values given in the Old and New Testaments and in the teachings of Buddah.

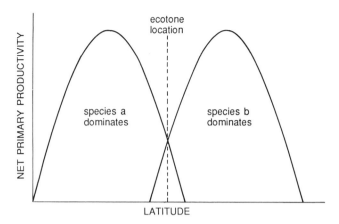

FIGURE 3.25 Hypothetical relationship of net primary productivity and latitude for two competing vegetation types. SOURCE: Reprinted, by permission, from Arris and Eagleson (1989). Copyright © 1989 by Ralph M. Parsons Laboratory, Massachusetts Institute of Technology.

the soil moisture supply. This amounts to maximization of soil moisture for given climate and soil.

Are these hypotheses valid? Are others preferred? Their verification would establish equations of state for equilibrium vegetation communities, thereby contributing significantly to solving such practical problems as optimal irrigation strategies, soil rebuilding, revegetation, estimation of effective mesoscale soil properties, recognition of changing climate, and many more. Once the appropriate equations of state are determined, the rates of change from one state to another can be addressed. This is of considerable concern in planning for and coping with the effects of climate change on soil-vegetation systems.

Microbial Transformations of Ground Water Contaminants

What is the nature of the feedback processes that occur between biochemical processes and the various physical transport mechanisms?

Subsurface microbiology and biotransformation of organic contaminants offer scientific and engineering challenges. Investigations that integrate both laboratory and field opportunities to make measure-

CHARLES R. HURSH
(1895-1988)

The Weeks Law of 1911 provided for the establishment of National Forests in the East, primarily to protect forest and water resources in headwaters of navigable streams. At the time, little was known about the influence of forests on water holding and regulation of streamflow. Interest in this issue was at a high level in 1926 when Charles R. Hursh, an ecologist, joined the five-man staff of the newly established Appalachian Forest Experiment Station. He pioneered research in ecology and forest hydrology, and many of the basic principles of wild land hydrology and watershed management are directly attributable to his innovative thinking and aggressive research leadership.

A native of Jonesboro, Illinois, he received a B.S. from the University of Missouri and a Ph.D. from the University of Minnesota. The first goal of his research was to define the characteristics of the soil, water, and climate of forests and abandoned agricultural land. He foresaw the need for complete instrumentation of watersheds to provide continuous measurements of precipitation, ground water levels, and streamflow. By 1932, Hursh had studied the various needs of the mountain and Piedmont regions of the Southeast and had prepared a comprehensive analysis of watershed problems and an approach to solving them. He concluded that "the purpose of the streamflow and erosion study in its broader sense is to determine the principles underlying the relation of forest and vegetative cover to the supply and distribution of meterological water." He also recognized the need for an interdisciplinary approach to the research. The indefatigable Hursh utilized manpower and funds from various federal relief programs of the 1930s (the Civilian Conservation Corps, the Work Projects Administration, and so on) to completely instrument components of the hydrologic cycle on numerous watersheds. He was small in stature, a giant in intellect, and completely intolerant of bureaucratic obstacles to science. During the late 1930s and early 1940s, Hursh's work in collaboration with other outstanding hydrologists, like E. F. Brater and M. D. Hoover, shaped current concepts of streamflow generation on forest land and quantified many effects of forest removal on streamflow and water quality. He developed the first successful infiltrometer and studied water movement through the soil profile. Although a strong advocate of the scientific method, Hursh was equally adept and successful in developing solutions to very practical questions. He was perhaps proudest of his research on naturalization of highway roadbanks to control erosion.

In later years, Hursh continued his advocacy for maintenance and improvement of environmental quality as a scientist-scholar and consulted in the United States and abroad, including France, Japan, Turkey, and Kenya. In 1953, he received the prestigious Nash Conservation

Award. His contributions to the hydrologic sciences are documented in over 125 publications. Perhaps the greatest tribute to his genius is that his work has stood the test of time—the research plan he conceived in 1931-1932 and implemented in 1933 became an internationally renowned hydrologic laboratory during his active career and, with little change in concept, continued as a world leader in hydrologic research for 50 years.

WETLANDS: KIDNEYS OF THE LANDSCAPE

Wetlands provide some of the most important ecosystems on the earth. At a global scale, wetlands represent about 6 percent of the total land surface, ranging from 17 percent in the subtropical zone to about 2 percent in the polar zone. Wetlands perform several valuable functions, serving as sources, sinks, and transformers of chemicals and biological materials; filters of polluted water; sources of ground water recharge; and buffers against flooding. Wetlands provide a unique habitat for a variety of flora and fauna. Historically, wetlands have been the centerpiece of cultural and economic development, including domestic wetlands such as rice paddies, which support a major segment of the world's population. For centuries, peatlands, salt marshes, and mangrove forests have all provided food and fiber essential to mankind.

Hydrologic processes are closely linked with chemical and physical aspects of wetlands, which in turn affect biotic functions and subsequent feedback that may alter wetland hydrology. Hydrologic processes directly influence species composition and richness, primarily production, organic matter turnover, and the cycling of nutrients in the wetlands. These highly productive ecosystems are frequently subjected to an annual flood period. The amplitude and regularity of flooding are of utmost importance in these fluctuating systems, where communities are adjusted to the pulse of seasonal variations in water levels. The rearrangement of hydrologic patterns through human intervention will result in large-scale ecological changes whose consequences cannot be well understood or predicted at our present level of understanding.

There is an urgent need for accelerated research on wetlands because of the tremendous development pressures these ecosystems face. Although efforts to protect wetlands have increased, they continue to be altered through drainage, dredging, impoundment, thermal and nutrient additions, filling, and conversion to agriculture. The value and the management of wetlands in their natural state hinge on basic hydrologic studies.

ments in contaminated aquifers are just beginning. The quantification of microbial effects on subsurface transport processes and the feedback mechanism of these processes on microbial ecology as well are still in their initial stages of development.

An example of such a feedback starts with a decrease in soil permeability caused by ion-exchange reactions (which brings about clay dispersion) or microbial action (which produces pore clogging by mucilaginous organics). In either case, the decrease in permeability decreases water fluxes. This, in turn, reduces the rate and extent of the original ion-exchange reactions or microbial activities. Will a steady state eventually be created? Or will transport stop entirely? Answers to these questions are of considerable interest in connection with soil formation, ore deposition, and pollutant migration, to name but a few applications.

Microbial populations can strongly influence rates of transport and the transformation of synthetic organic compounds. The dynamics of this phenomenon are not yet well understood, especially at the field scale. Nevertheless, it is clear that they are annually dependent on the growth mechanisms of the microbe colonies, which are themselves linked to complex hydrologic and geochemical interactions. There is a need to develop basic understanding of this complex system, including the relationships among its parameterizations at different scales of observation.

HYDROLOGY AND CHEMICAL PROCESSES

Introduction

Chemical processes in the hydrosphere determine the composition of natural waters, control rates of chemical weathering and the geochemical cycles of most elements, and influence the chemistry of both the earth's crust and its atmosphere. Trace atmospheric gases emanate in large part from aquatic and wetland systems, influencing climate and hence the hydrologic cycle itself. Finally, the chemical composition of the water, as much as its quantity, determines its ability to support life. **Chemical and biological processes in water are tightly intertwined, influencing each other and determining not only the "quality" of the water but also the nature of life on the planet.**

Understanding the interactions of chemical and hydrologic processes by which the chemistry of the earth's waters and atmosphere is shaped is no longer a matter of purely academic concern. Today, the very ability of our planet to sustain a stable environment for life is challenged by human interventions, chemical and hydrologic, on a global scale. The dramatic effect of these chemicals on the water quality of streams can be seen in Figure 3.26, which shows a nearly tenfold increase in the flux of chloride during the last century for the Rhine River. Hydrologists, chemists, and biologists face the challenging task of understanding earth processes that inseparably involve concepts from each discipline.

In addition to demanding interdisciplinary synthesis, the frontier questions of environmental chemistry require additional emphases within the traditional disciplines themselves. Hydrologists will need to focus more on the actual pathways and residence times of individual water parcels in different environmental compartments. Although many traditional problems in hydrology (e.g., large-scale water balances, flood forecasting, and ground water resource evaluation) do not require such knowledge, chemical considerations demand knowledge of water flow paths and hydraulic residence times. Knowing where a parcel of water has been, and for how long, often is the key to understanding its chemical evolution.

Recognition and incorporation of hydrology into conceptual models of natural water chemistry will require environmental chemists to advance the state of existing chemical kinetic models. Historically, the dominant paradigm of aquatic chemistry—the study of chemical

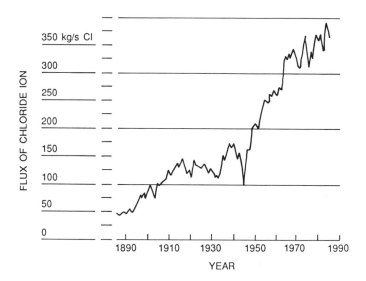

FIGURE 3.26 Changing chloride flux in the Rhine River at the border of The Netherlands and Germany. SOURCE: Personal communication from T. S. Uiterkamp, 1985 Annual Report of the Cooperating Rhine and Meuse Drinking Water Companies.

processes in natural waters—has been that of chemical equilibrium. Because most reactions were considered very fast (e.g., acid-base reactions over microseconds) or very slow (e.g., geological alteration of minerals over millennia) compared to human time scales, the historic focus has been on (pseudo) equilibrium presumably achieved among the many components of aquatic systems over periods of hours to years. This convenient paradigm must now be reconciled with the realization that the time scales of many important chemical processes are in fact within the range of pertinent hydrologic residence times. Some reactions that were thought to be fast have turned out to take hours; some otherwise slow reactions are in fact accelerated by unforeseen catalytic, photochemical, or biochemical mechanisms.

In summary, the extent of chemical reaction in environmental water is determined by the chemical rates of reaction relative to the hydrologic residence times, and these quantities can be of similar magnitude. The need for new environmental syntheses, drawing upon and expanding the scope of both hydrology and chemistry, is abundantly clear.

Some Frontier Topics

Effects of Acid Rain

The 1980s saw a rise of public awareness that industrial society is beginning to influence the earth's chemistry on a regional and global scale. The "acid rain" issue is a dramatic example. Surface water acidity and associated toxicity to fish have led the public's list of concerns. More recently, concern has also grown about the nutrient effects of nitric acid deposition and possible resulting eutrophication of natural waters (e.g., EDF, 1988). Understanding hydrology is central to understanding and predicting acid rain effects. At every step along the pathway of water, from the formation of precipitation to the eventual discharge to the oceans, crucial and challenging questions remain.

Precipitation Chemistry

Although a broad understanding of the factors that determine the chemical composition of rain and other atmospheric deposition exists, the available data base limits the development of truly quantitative approaches. Research using rainfall composition data as tracers of the hydrologic cycle offers opportunities to understand better the relationship between rainfall composition and the chemistry of soils, surface water, and ground water.

What are the mechanisms whereby atmospheric pollutants are incorporated into rain?

Rain is not pure, distilled water. Both rain and snow contain a wide range of dissolved and suspended materials, both organic and inorganic. However, it has only been during the last two decades that analytical facilities have been capable of demonstrating the great variety and range in concentration of substances included in precipitation.

Gases, liquid droplets, and particulates are released into the atmosphere from natural and anthropogenic sources on the land surface. These emissions differ widely in composition from place to place and with time. During atmospheric transport, gases and particles interact physically and chemically under the influence of radiation, temperature, and humidity, yielding new compounds. At any particular site, the composition of the atmosphere will reflect several factors, including the direction of the wind, the character of the sources, the extent of mixing from multiple sources, and the chemical reactions that have occurred. When snow forms in the atmosphere, it can incorporate particles and gases, some of which are soluble in water. During descent, if temperatures are high enough, the snow will change to rain. Rain selectively scavenges both gases and particles from the air as it falls, and becomes enriched in these materials to a degree that will vary with rainfall intensity, chemistry of the dissolved material, and history of the air mass. If snow instead accumulates on the ground, it will serve as a substrate for dry deposition from the atmosphere, and, upon melting, will dissolve a part of these deposits.

Recognition of acid rain as a problem in Europe and North America has indeed greatly stimulated interest in the chemistry of atmospheric deposition. This interest creates opportunities not only for solving practical problems, but also for advancing understanding of the basic chemical processes. There has been a major increase in research on the sources and transport of the chemicals removed from the atmosphere by rainfall, as well as investigations of the mechanisms whereby atmospheric pollutants are incorporated into rain and snow. That research, however, has not yet produced quantitative models with which the composition of wet deposition can be predicted on the basis of the factors noted above.

ACID PRECIPITATION

What is the scientific problem?

Strong acids from atmospheric deposition can significantly affect the natural process of soil development. There are many acid-sensitive catchments, generally underlain by granitic bedrock and with shallow, acidic soil with small pools of labile, easily weatherable metals like calcium or potassium. These catchments may also have little ability to adsorb strongly acidic anions (sulfate, nitrate) on soil minerals or organic compounds; such catchments are said to have low acid-neutralizing capacity. Therefore acids are not neutralized in the soil but can react with aluminum-bearing minerals or be flushed into streams and lakes. These acidic cations of hydrogen and aluminum are toxic to aquatic life. Areas at risk from this degradation of surface water quality include the northeastern United States and Canada, Scandinavia, and mountain ranges such as the Sierra Nevada and the Rocky Mountains.

There is potential here for damage to human health as well. Acidification of water supplies brings increased concentration of potentially toxic metals such as lead, cadmium, mercury, and aluminum in that water, leached from soils, sediments, and pipes and fixtures in the water supply systems. Aluminum, which makes up 5 percent of the earth's crust, is practically insoluble in neutral or alkaline waters and thus not historically available to animals and humans. Within the last decade, however, high concentrations of aluminum have been found in tissues of patients who have undergone long-term dialysis in areas with significant aluminum in the water. High concentrations of aluminum have been found in the brain tissue of patients with Alzheimer's disease and with senile dementia. Whether or not there proves to be a causal relation between these examples and water quality, the threat of unsafe drinking water adds motivation to understand the hydrologic process of acid precipitation and its role in the chemical and nutrient balances of drainage basins.

What is the role of the hydrologic sciences?

Acids from atmospheric deposition may have altered the natural process of soil development principally by facilitating the transport of weatherable aluminum from soil to surface waters. Evidence to support this contention is available in data on water and soil chemistry collected from acid-sensitive regions that are receiving a high atmospheric loading of acids, such as the Adirondack region of New York and New Hampshire. Organic soil horizons there contain remarkably high concentrations of aluminum, which appeared to be bound mostly with humified organic matter. Solutions from soils supporting coniferous vegetation contained higher aluminum concentrations in all soil horizons than those from soils supporting deciduous trees.

These observations suggest that vegetation assimilation and microbial decomposition may be an important component of the aluminum bio-geochemical cycle in forested ecosystems.

What is the next step?

Better knowledge is needed of the flow paths of dissolved and particulate aluminum under different precipitation rates, of the chemical character of aluminum in soils, and of the factors that determine its fate. The rate at which the aluminum pool in acid-sensitive catchments forms and the factors influencing the rate are unknown.

Reliable procedures are needed for measuring the chemical forms of aluminum in natural waters. Inconsistent findings about the magnitudes of the processes regulating aluminum concentrations in runoff are partly the results of researchers relying on measurements of total aluminum concentration.

Snowpack Chemistry

What are the measurement techniques and modeling schemes necessary to better understand the spatial and temporal distribution of the rates of chemical release from snow throughout a catchment?

Snowflakes may form around microscopic pollutant particles in the atmosphere. Once they have served to initiate the condensation process, these particles become encased in the growing ice crystal. As the snowflakes fall, they may scavenge pollutants from the atmosphere. Once they form a snow cover, they generally receive a continuous input of pollutants by dry deposition. The strengths of these inputs can be so great that, in some locations, it is possible to tell the direction of air movement by the color of the snow: Saharan dust reaches the snow-covered Alps, and the Scottish mountains can be tinted black from coal burning in England.

As the seasonal snow cover thickens with the included impurities, the snow undergoes continuous metamorphism (Figure 3.27), which has been hypothesized to move the impurities to crystal surfaces. This phenomenon is believed to occur because of the continuous grain growth that takes place in snow and the constant exchange of ice mass from grain to grain as water vapor moves down the temperature gradient in dry snow. As ice crystals grow, they should reject almost all solute impurities; thus snow grains undergo purification from the onset of melt.

The availability of impurities on the grain surfaces contributes to "acid shock," which is observed commonly now but was first identified as a major problem only in 1978 (Johannessen and Hendriksen, 1978). The initial fraction of meltwater causes the sudden removal of a large fraction of the acid-producing solutes that have accumulated in snow over the winter months. This causes a sudden surge of polluted meltwater as a direct result of the concentration of impurities on the grain surfaces during the metamorphism described above. Events during winter such as midseason melt can concentrate impurities near the bottom of snow cover to produce more rapid removal. Thus the events during any particular winter can have a considerable effect on the details of an acid flush. Large variations in the chemical contamination of snow occur vertically over the scale of the thickness of snow cover and laterally over the regional scale. Many data, gathered under demanding conditions, are necessary to provide a clear picture of the level of contamination and the mobility of the contaminants.

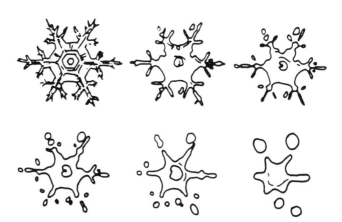

FIGURE 3.27 Snowflakes are destroyed rapidly by metamorphism. In the process, contaminants, such as acids, are redistributed and concentrated in a complex fashion. SOURCE: Reprinted from Bader et al. (1939) courtesy of U.S. Snow, Ice, and Permafrost Research Establishment.

Because of the wide range of weather patterns in different areas and in different years, all of the possibilities are not well understood, and the consequences of increasing acid deposition in areas that are now relatively acid-free are not yet predictable. These uncertainties arise in part because of large variations that occur in snow accumulation and melt rate, even in small alpine catchments. The spatial and temporal distribution of the rates of chemical release from snow throughout a catchment must be evaluated. This evaluation will require data on the initial distribution of chemicals and their elution from the snowpack in response to melt and metamorphism. Changes in elevation and in aspect produce such large changes in energy budget factors (such as solar radiation) that different parts of a catchment often melt out during different parts of a season. The effect of the acidic meltwater also varies spatially over a watershed (e.g., poor buffering in relatively unweathered zones and the release of trace metals from soils). Even if meltwater penetration into lakes produces little effect (for example, if the lakes are well stratified), the resulting flush of acidic meltwater down their outlet streams may not be acceptable.

Fate of Acid Deposition in the Soil Environment

What is the relative importance of different flow paths and residence times to the chemistry of subsurface water?

Although rain and meltwater may in part enter a stream directly, in many areas most of the water enters the soil environment prior to emerging as streamflow. Subsurface flow pathways bring water into contact with several effective agents of chemical change. Quantitative ignorance of these pathways is a major obstacle to understanding the chemistry of the water when it re-emerges as streamflow.

Weathering, microbial transformations, production of humic materials, and plant uptake profoundly influence subsurface water chemistry; yet, models of each are in their infancy. Moreover, their relative importance is governed as much by hydrologic flow paths as by purely chemical and biological factors.

Figure 3.28 diagrams the interrelated physicochemical phenomena that combine to control acidity in the soil water (the soil solution). In addition to rain and melted snow, the physical inputs can include fog and dry deposition (particles and acidic gases such as sulfur dioxide and nitric acid vapor). Deposition comes both from direct atmospheric

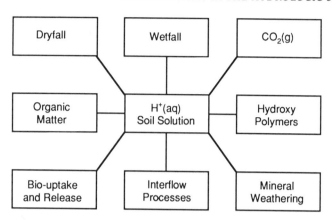

FIGURE 3.28 Some physicochemical factors controlling the acidity of water in soils. The close interaction of both chemical and hydrologic processes is involved in the ultimate determination of soil water chemistry. SOURCE: Reprinted, by permission, from Sposito (1989b). Copyright © 1989 by Oxford University Press.

inputs and from throughfall below the vegetation canopy, and measurements must be made on time scales that are relevant to the runoff cycle and on space scales that reflect catchment physiography. These same considerations apply to acid-exporting processes (e.g., volatilization, eolian transport, and streamflow).

The important chemical phenomena that influence the acidity in the soil water are (1) acid-base reactions involving carbonic acid, soil humus, and mineral weathering products (e.g., aluminum-hydroxy polymers); (2) ion-exchange and mineral weathering reactions; (3) vegetational uptake and release of ions; and (4) microbiologically mediated processes (such as nitrate immobilization and sulfate reduction).

Although carbon dioxide concentrations in the soil atmosphere exceed those in the open atmosphere by more than 1 order of magnitude, and soil humus is the single most important repository of acidity in soil, almost no quantitative, catchment-scale data exist on soil carbon dioxide characteristics, on the rates of production and decomposition of humus, and on humus-mediated effects on soil acidity.

Mineral weathering rates, typically investigated in the field through geochemical mass balance techniques, range over at least an order of magnitude for catchments in the United States, and their functional relationship to microscale chemical kinetics and mechanisms is essentially unknown. Extrapolation of the existing data base to unstudied catchments is not possible, particularly in light of the oversimplified models used to interpret geochemical mass balance data (e.g., idealized mineralogy, constant reaction stoichiometries, and constant rates of transformation).

Biological processes important to the regulation of acidity in soils include nutrient uptake and release and the microbial mediation of chemical reactions. The soil environment near plant roots can become either acidified or more alkaline, depending on the overall nutrient uptake and release characteristics of the plant. Fluctuations in soil pH near roots of as much as two pH units have been observed. These processes are not well characterized, especially at the catchment scale. Similarly, the effect of microbial populations on chemical reactions that consume or produce acidity is not known on the catchment scale. Further, the implications of the perennial but gradually changing nature of forest biota have not been worked out.

Perhaps most importantly of all, the flow paths and residence times of water in this complex subsurface environment are not known, despite the overriding importance of knowing in which soil layers or microhabitats, and for how long, the soil water resides. This knowledge gap is especially evident in current acid-rain-effect models, which are deemed important to the eventual formulation of national acid rain regulatory policy. In such models, for lack of knowledge the water flow paths are represented arbitrarily or are omitted entirely (i.e., a whole catchment may be treated as if it were a fully mixed vessel of soil particles and water). There is a pressing need for closer integration of hydrology with environmental chemistry, as well as for advances in the tools used to map out water flow pathways in the soil environment.

Contaminant Fate and Transport

The frontier issues in ground water and soil water contaminant fate and transport share much with the issues of subsurface acid deposition effects described earlier. Usually, however, the chemicals of concern issue from point sources. Many are toxic, and the hydrogeologic setting is often that of an economically valued aquifer used for water supply. One important frontier area encompasses the problems of heterogeneity at several scales. Others involve the complexity of interaction between reactive solutes and soil or aquifer solids, the possible transport of very small particles (colloids), and the effects of microbial communities on both contaminant fate and subsurface permeability.

These frontier problems are discussed in more detail in a previous section, "Hydrology and the Earth's Crust." They bear reiteration as a set of issues that is of particular relevance to current concerns about hazardous waste in the environment. Added to these issues is the critical role of the biogeochemical cycles that couple terrestrial eco-systems to subsurface runoff flow paths. The mechanisms, mass transfer rates, and mass storage capacities in these flow paths and cycles vary with the chemical element concerned and exhibit significant spatial and temporal variability within and among catchments. Their key importance to water quality management derives from the natural pathways of detoxification they can provide for potentially hazardous chemical compounds deposited on land. Better knowledge of the flow paths of water under differing water loading rates, of the chemical character of the labile pool of a compound in soils, and of the factors influencing bioturnover rate are necessary before we can predict the quality of receiving waters to be managed for beneficial human use.

Sediments also play an important and incompletely understood role in the behavior of chemicals in surface waters. Sediments range widely in grain size, mineralogy, surface area, organic matter content, extent of chemical reactivity, and rate of reaction with surface water solutes. Some move in suspension, whereas others are transported slowly over the streambed or are temporarily immobile in stream bottom deposits. Sediments may control the concentration of a solute dissolved in water or may have little effect on water quality, depending on sediment reactivity, the extent of sediment-water contact, and the chemical characteristics of the solute. For example, hydrous oxides of iron and manganese in sediments bind toxic metals (e.g., lead). Solid organic matter can react strongly, not only with organic solutes, but also with inorganic ions. Clay minerals, with their small size and large surface areas, serve commonly as substrates for deposits of various substances (e.g., iron and manganese oxides, as well as organic matter). Such deposition greatly enhances the availability of these oxides and organic materials for subsequent reaction with dissolved surface water solutes. Clay minerals themselves carry adsorbed ions that can be exchanged with other ions in solution, thereby affecting water quality. Thus one must understand sediment-solution interactions in order to understand the chemistry of surface waters.

What is the relative rate of transport of various sizes of polluted aggregates and grains in stream waters, and how long are such particles stored at various locations before moving on in the channel system?

> How much transfer of adsorbed materials from one grain to another occurs during streambed storage, with differing interbed flow and dissolved oxygen availability?

The degree of affinity between a solute and sediments can vary greatly, depending on the nature of the solute and the characteristics of the sediment. Some solutes, including many toxic metals, like mercury, can be adsorbed very strongly by sediments. Others, like sodium, are held weakly. Changing either the concentration or the chemical form of the adsorbing solute, or altering the composition of a surface water, can alter radically the affinity with which a solute is held. Thus, if clean water flushes a polluted stream, some release of pollutants from the sediments will occur even though the pollutants are adsorbed strongly. If the stream discharge also increases, the bed sediments may be disturbed, allowing resuspension of interstitial fine particles into the water column and migration of coarser particles as bed load. Alternatively, if labile organic matter and sediments containing toxic metals bound in iron and manganese coatings are deposited together on the bed under quiescent conditions, an oxygen-depleted, reducing environment can develop, causing dissolution of the iron and manganese oxides and release of these metals into the stream water. Organic pollutants can also be changed during storage in the bed, and, when the next stream rise occurs, they may be transported in an altered chemical form.

Sediments derived from commonly occurring landforms can be expected to contain fine-grained aggregates ranging from silt size up to boulder size. Chemical reactions between these aggregates and solutes are influenced not only by the transport of reacting solutes to the exterior surface of the aggregate (film diffusion) and the reaction time (often very short), but also by the time needed for the participating ions to diffuse into and out of the aggregates (particle diffusion). This means that a surface water system can develop a significant "memory"; therefore a common simplifying assumption, that adsorbed solutes are transported almost entirely in dispersed fine-grained sediments, can be very wrong.

The importance of sediments in concentrating chemical elements from solution is also recognized. When aquatic organisms ingest polluted sediments, the pollutants can become incorporated into tissues, and, as smaller organisms are consumed by larger organisms along the food chain, the pollutants can be concentrated to the point where human beings, the ultimate consumers, can be harmed. These and many other issues remain to be addressed before a thorough under-standing can be achieved of how specific solids and solutes interact and move through surface water systems. There are many opportunities to gain an improved basic knowledge of the operation of natural water systems while contributing to solving a major problem for an industrial society—that of water quality.

Global Chemical Cycles

What are the feedback links that control the interaction between the hydrologic cycle and the global chemical cycles of crucial elements, such as carbon and nitrogen?

Several elements have biogeochemical cycles that are important at the planetary scale and are linked to the habitability of the planet as well as to the distribution and productivity of its ecosystems.

The geochemistry of carbon provides a clear example of the im-portance of global elemental cycles in hydrology. As greenhouse gases in the stratosphere, carbon dioxide and methane are of great interest to climatologists and hydrologists. Carbon dioxide is also the dominant natural acid in the hydrosphere, and its concentration in water influences the weathering of minerals. In addition, fixation of carbon dioxide by photosynthesis and its regeneration by respira-tion are among the very principles of life on the earth. Thus major processes in the atmosphere, hydrosphere, lithosphere, and biosphere are linked, influencing each other through the geochemical cycle of carbon. Ultimately, our general circulation models of the atmosphere will have to be interfaced not only with ocean circulation models (more physics), but also with geochemical models of the cycles of key elements such as carbon (chemistry).

Nitrogen is a second element whose cycling is intimately linked to hydrology and whose movement and transformation profoundly influence life on the earth. Biologically available nitrogen (e.g., nitrogen as ammonium or nitrate) is a major required nutrient for all life. Its availability is believed by many to limit the production of the oceans as well as of many terrestrial systems; its excess can initiate eutrophication and ecosystem degradation (as in the Chesapeake Bay). In its gaseous forms, nitrogen as nitrogen dioxide is an ingredient in the production of photochemical smog and health-threatening ground-level ozone, and nitrogen as nitrous oxide both attacks the stratospheric ozone shield and contributes to the greenhouse effect.

The major biological transformations of nitrogen occur in, or are closely tied to, soil water and surface water systems. The hydrologic cycle transports biologically available nitrogen, as well as determines conditions (oxic or anoxic, long or short water residence times, moist or dry) that control the transformations. As with carbon, the global cycles of this element are intimately linked with hydrology.

Several of the radiatively active gases implicated in global climate change are produced in anoxic systems whose existence hinges on the extent of water saturation. Our present ability to predict the response of this hydrologic-chemical-biological system to perturbations is almost nil. An example is the problem of methane release from wetlands.

Methane gas follows water vapor and carbon dioxide in importance as a greenhouse gas. Like carbon dioxide, its presence in the atmosphere is growing, its concentration increasing at a rate of about 1.7 percent per year (Figure 3.29). Major sources are believed to include combustion,

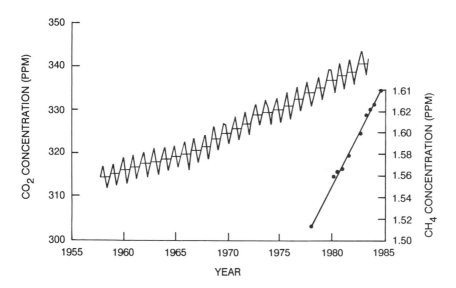

FIGURE 3.29 Worldwide concentrations of tropospheric carbon dioxide (CO_2), sampled from 1958 through 1984 at the Mauna Loa Observatory, Hawaii, and of methane (CH_4), derived from simultaneous Northern and Southern Hemisphere samples of surface air from 1978 through 1984. Units of measurement are parts per million, by volume. Methane and carbon dioxide are both important greenhouse gases, and both are intrinsic in biospheric processes. SOURCE: Carbon dioxide data courtesy of C. D. Keeling, Scripps Institution of Oceanography; methane data courtesy of D. R. Blake, NASA/National Space Technology Laboratories, and F. S. Rowland, University of California at Irvine (UCAR, 1985). Published by University Corporation for Atmospheric Research, Boulder, Colorado.

fossil deposits, the rumen of cattle, and wet ecosystems such as rice paddies and natural wetlands. In wetlands, methane is produced in waterlogged environments but in close proximity to the atmosphere, thus facilitating release to the air. The complexity of methane production and release by wetlands is illustrated by Figure 3.30. Vegetation and waterlogged sediment together provide oxygen-depleted conditions that permit the generation of methane by microbes. The same sediment controls the export of methane (which can occur by dispersion, convection, and bubbling), in part by harboring methane-consuming microbes near the surface.

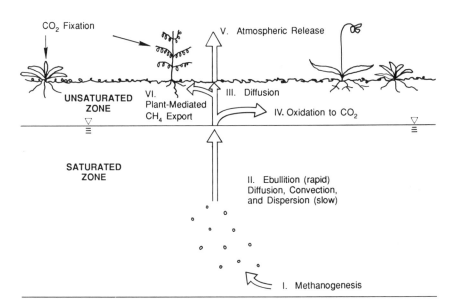

FIGURE 3.30 Processes involved in methane evolution from northern peatlands. Methane produced by bacteria (I) may be transported toward the surface by several physical processes (II). Methane reaching the unsaturated zone faces competing fates of reoxidation (IV) or physical transport (III) to the free atmosphere (V); plants (VI) may also transport methane upward. Hydrology directly affects each step of the process.

ROBERT M. GARRELS
(1916-1988)

Robert M. Garrels pioneered and dominated the modern scientific study of the chemistry of natural waters. He also reinvented the notion of the geochemical cycle. Both areas have their roots in the earliest days of the natural sciences and attracted the attention of the great polymaths, both chemists and physicists, of those times. However, as the straitjacket of academic specialization steadily tightened, geologists came to know less and less physics and chemistry, and physicists and chemists learned no geology. Garrels reachieved the ancient synthesis placing these fields firmly in the scientific mainstream.

Garrels was born in Detroit and received degrees in geology from the University of Michigan and Northwestern University. He held positions at Northwestern, the U.S. Geological Survey, Harvard University, Scripps Institution of Oceanography, the University of Hawaii, Yale University, and the University of South Florida. His interest in the application of thermodynamics to geologic processes was first stimulated by the study of ore deposits, particularly roll-front uranium ores. In a classic paper he showed that these were precipitated from flowing ground water by changes in redox conditions. He than generalized this result in a monumental text, *Solutions, Minerals, and Equilibria*. The redox diagrams in this book achieved such complexity that the construction of several of them earned M.S. degrees for his Harvard students.

Garrels then extended this approach from oxides and sulfides to the aluminosilicates. He developed equilibrium models for weathering reactions and, extending the ideas of the great solution chemist L. G. Sillen, produced a beautiful model of the processes that control the composition of seawater. This effort culminated in a book, *Evolution of Sedimentary Rocks*, that founded a veritable industry based on modeling the geochemical cycle of the elements over geologic time.

In his later years, Garrels became something of a national institution, the patron of a "Flying University" devoted to the study of environmental geochemistry. His enthusiasm and wit attracted acolytes of all ages, and his mastery of the Socratic method enveloped them in the problems uppermost in his mind. As befitted the son of an Olympic medalist (1908, in shotput and hurdles), he was active in many sports, exhausting many colleagues. He left an imprint on his field such as few others can claim.

Hydrologic activity profoundly influences each step of the overall process. Water movement determines both convective and dispersive methane movement. Water table height controls the proportion of organic sediment that is saturated, anoxic, and methane-producing, relative to the upper, unsaturated, methane-consuming layer. Small (tens of centimeters) changes in water table height can shift a wetland from a methane-evolving to a methane-consuming system. Potentially sensitive feedback links may thus exist, perhaps on a global scale, whereby methane evolution helps drive climate change, which in turn, via hydrologic effects, alters methane evolution.

To understand such a complex system will require the full integration of chemical, hydrologic, and biological disciplines. This frontier task has not yet been seriously undertaken and is one of several important challenges facing the field of hydrology in the next decades.

HYDROLOGY AND APPLIED MATHEMATICS

Introduction

Experimentation/observation and theorization are the two pillars on which much of modern science and technology rest. **Mathematics provides the fundamental logical framework for developing systematic and logically consistent theories of physical phenomena.** The early attempts at a quantitative understanding of hydrologic phenomena were directed toward laboratory experimentation of individual processes, for example, water movement over land, in channels, and into and beneath the surface, and their mathematical characterizations. These efforts led to a fairly comprehensive laboratory-based understanding of many individual components of the hydrologic cycle.

At larger space and time scales, controlled experimentation becomes more difficult, and at the largest scales, all but impossible. Then nature itself has to be treated as a laboratory for taking observations of the natural phenomena. However, in a natural laboratory it is impractical to isolate and investigate individual effects on a given phenomenon. So one must frequently be content to observe the physical phenomenon at hand rather than all the individual processes that govern that phenomenon. Owing to the highly nonlinear nature of hydrologic processes, and their mutual interactions, typical observations in space and time show a high degree of variability over a very wide range of space and time scales. A major challenge currently facing hydrologic science is to uncover the physical message of observations at space and time scales other than the laboratory scale and develop logically consistent and testable mathematical theories of hydrologic phenomena.

Specifically, theoretical studies are undertaken for identifying and formulating general mathematical constructs and providing testable theories for the purposes of (1) providing theoretical interpretations and explanations of empirically observed regularities and structures, (2) unifying disjoint empirical relationships and/or processes across scales, (3) discovering hidden order in physical phenomena, e.g., symmetries, and invariance across scales, that specific models must obey, (4) providing guidelines for new measurements and experimentation for testing general, but important, mathematical hypotheses, and (5) providing qualitative theoretical insights to help in understanding the key features of a physical phenomenon and in making predictions.

Recent advances in the mathematics of probability and stochastic processes and in nonlinear equations and numerical methods and major growth in computing capabilities and in remote sensing technology all point toward new and exciting opportunities in the development of hydrologic theories over a broad range of space and time scales. In this section, some of the emerging new directions in theoretical investigations of hydrologic phenomena are described to illustrate the ideas and objectives listed above.

Some Frontier Topics

Scaling and Multiscaling Invariance in Spatial Variability of Hydrologic Processes

"This paper is an invitation for the reader to solve the problems of pure and applied geometry involved in its approach to the notion of length and shape rather than an attempt of the author to answer the questions by himself." So reads the abstract to the classic paper by Hugo Steinhaus (1954) entitled "Length, Shape, and Area." The problems of "pure and applied geometry" that had captured the attention of one of this century's most eminent mathematicians centered on a hydrology problem. Specifically, Steinhaus sought to determine a practical means of assigning "length" to rivers described by maps drawn at different scales under the full recognition of the fact that rivers have no intrinsic length scale. A contemporary statement of this problem would begin with the recognition that the river is a naturally occurring "fractal," or exemplifies the "paradox of length," in the words of Steinhaus, and then ask how theoretically meaningful measurements of such structures can be had. The problem has a depth that makes the invitation no less important to hydrologic interests today than it was some 35 years ago.

So how does one contend with the "issues of scale" that are so pervasive in hydrologic phenomena? For the river problem, Steinhaus offered a probabilistic method that continues to hold the promise of a foundation for revolutionary new approaches to measurement. Another perspective is illustrated by the related work of Albert Einstein (1926) in his classic paper entitled "The Cause of the Formation of Meanders in the Courses of Rivers and of the So-called Bear's Law." This work holds the fundamental view that, in spite of the fractal paradox of length, there are scales on which regularity may be both measured and physically explained. This view is explored by Einstein in his qualitative physical explanation of the cause of formation of meanders and of the observation that meander frequency decreases with an increase in river cross section.

The interplay between probability, physics, and geometry is by now deeply rooted in hydrologic theories and practice. While stochastic approaches to the analysis of average properties of single rivers are prominent in the work done by Luna B. Leopold, Walter B. Langbein, and others in hydrology in the 1950s and 1960s, the search for the next step is at the very frontier of contemporary science and mathematics (Leopold and Langbein, 1962). And as one moves from the case of single rivers to that of the multichanneled branching networks found in river basins, the interplay becomes all the more exciting.

What are the scaling and multiscaling properties of three-dimensional drainage networks?

In his paper entitled "Erosional Development of Streams and Their Drainage Basins: Hydrophysical Approach to Qualitative Morphology," Robert E. Horton (1945) devised an ordering scheme for encoding and measuring the degree of complexity of bifurcations in river networks that displayed observable patterns of pronounced regularity. These observations influenced in no small way the developments of quantitative-empirical geomorphology over the next two decades. However, it was not until the late 1960s that Horton's observations were interpreted statistically by R. L. Shreve in a series of theoretical papers. The manner of the statistical interpretation leads once again to unresolved issues of deep physical, mathematical, and practical importance. For example, Leopold recognized the simple utility of Horton's law of stream numbers for estimating total lengths of rivers from maps; based on a simple geometric progression uncovered by Horton, the estimated total length of rivers in the United States is roughly 2 million to 3 million miles (Leopold, 1962). However, the accuracy of such an estimate is largely unknown. Problems pertaining to rigorously establishing laws of averages and the determination of error bars (central limit problems) for such estimates represent new mathematical territory that evidently holds significant new results for both hydrology and probability theory. Moreover, the extensions of the statistical theory to hydraulic-geometric quantities, such as river slopes, widths, depths, flows, and sediments, in networks involve the very essence of the scale issues embedded in the works of Steinhaus and Einstein.

Scaling methods have been and continue to be developed and explored in physical-statistical theories to describe properties of matter at phase transition, turbulence in the atmosphere, unstable biological populations, the Hurst effect in river flow statistics, and so on. An example that conveys the depth of such approaches is furnished by the following modern approach to the classical problem of Holtsmark. Specifically, one considers a random suspension of electrical charges in the atmosphere and asks for the probability distribution of the force exerted on a unit charge located at a fixed point in three-dimensional space. Now, doubling, tripling, and so on, the density ρ (charge per unit volume) corresponds to rescaling the length ℓ by an inverse cube root of two, three, and so on. If the Coulomb inverse square force law of classical physics is assumed, then the probability distribution of force at a given scale of ℓ units is standardized under rescaling by $\ell^{2/3}$. The statistical scaling exponent $2/3$ combines the physics of inverse square law with the geometry of space dimension. If the locations of the charges are taken to be statistically independent, then the probability distribution of the force is uniquely determined by the exponent $2/3$. Such is the potential for modern scaling methods in physical theories. **The search for an invariance property across scales as a basic hidden order in hydrologic phenomena, to guide development of specific models and new efforts in measurements, is one of the main themes of hydrologic science.**

Modern approaches to problems in hydrology are moving toward scaling theories as much out of pragmatic necessity as out of pure scientific curiosity and rigor. This is true whether one looks at current theoretical efforts dealing with water in the atmosphere, on the surface, or beneath the ground. The pragmatic reason is that hydrologic understanding and predictions are needed over a broad range of scales, ranging from 100 m to 10,000 km in space and from a few minutes to many years in time. Over such a range, measurements are hard to make and hard to follow because of noise and nonlinearity. Therefore

it is all the more important to make theoretically meaningful observations on such natural systems that are subject to the paradox of
measurement. The purely scientific reason happens to be the same
one. If the spatial and/or temporal variabilities embody a fundamental
hidden order that manifests itself across a wide range of scales as an
invariance property, then it must be formulated mathematically and
tested empirically, for the presence of such a property must be obeyed
by more specific mathematical models.

Recent investigations of the scaling properties of river networks
have provided new theoretical insights into the widely known empirical
feature that, on the average, rivers drained by larger drainage basins
have flatter slopes than those drained by smaller basins. Specifically,
the empirical mathematical relationship characterizing this feature
has been shown to follow as a consequence of the simple scaling
invariance property of river slopes in a channel network (see the
earlier discussion in this chapter's section titled "Hydrology and
Landforms"). The first-order predictions of the theory based on simple
scaling are remarkably close to empirical observations, but data quite
clearly suggest the need for generalizations. A promising avenue,
called multiscaling, is currently being explored for such generalizations.

While simple scaling properties of stochastic processes have a relatively
long tradition in the more advanced mathematical theories, multiscaling
processes are only beginning to be understood mathematically, mostly
within the context of statistical theories of turbulence. In simple
terms, the idea is partly that simple scaling implies a log-log linearity
in the plot of any statistical moment versus the scale of measurement
together with a linear relation between the order of the moment and
the slope of the log-log relationship. Figure 3.31 illustrates this feature
schematically. This feature leads to the interpretation that the statistical
spatial variability in the physical process does not change with a
change in scale; simple scaling processes are therefore said to possess
the self-similarity property. However, preliminary empirical analyses
of spatial rainfall, river flows, and channel gradients in networks
suggest that while the log-log linearity is preserved between the moments
and the scales of measurement, the corresponding slopes exhibit a
nonlinear concave growth with the order of the moments. In contrast
to linear slope behavior in simple scaling, the nonlinear concave growth
of slopes shown in Figure 3.31 leads to the interpretation that statistical spatial variability in such processes increases with a decrease in
spatial scale. In this case the scaling behavior is determined by a
spectrum of exponents, as opposed to a single exponent in simple
scaling; hence the term "multiscaling." Determination of these exponents
in terms of measurable physical and geometrical parameters of river

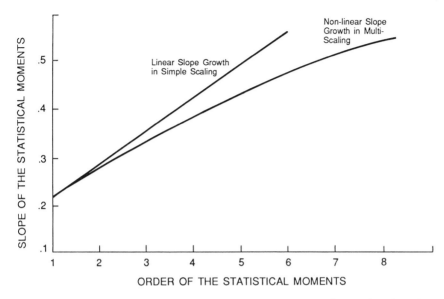

FIGURE 3.31 A schematic display of the difference between scaling and multiscaling behavior of empirical data.

basins is a problem of great significance. Multiscaling seems to arise in those physical systems that are governed by highly nonlinear dynamics.

Clearly, tests of a multiscaling invariance property will require new, more extensive measurements than are generally available at present. On the application side, such theories will guide hydrologic predictions for river basins under sparse data sets, and due to anthropogenic changes, as well as aid in the development of fundamental mathematical models of the hydrologic phenomena.

Stochastic-Dynamical Analysis of Hydrologic Time Series

An important task in hydrology—or in any other branch of natural science, for that matter—is to reconstitute from the data the principal features of the underlying physical system and to make predictions about its future evolution. Now, a typical time series pertaining to hydrologic data displays considerable complexity, without any obvious periodicity and with random-looking excursions of the relevant variables from their average level. A question of primary concern, therefore, is how to decipher the message of such a time series and thus understand the role of systematic effects and randomness from a physical perspective.

When confronted with a complicated, erratic succession of events, the first issue that comes to mind is that the phenomenon of interest is blurred by the presence of a great number of variables and poorly known parameters, whose inevitable presence is hiding some fundamentally simple underlying regularities. Traditional statistical methods provide useful algorithms for extracting the relevant signal from what is believed to be parasitic, random noise. Nevertheless, the traditional statistical analyses are not always telling the whole story, and it is important to adopt complementary approaches that are based on a more dynamical view of the hydrologic phenomena.

The importance of a dynamical approach stems from recent developments in physical and mathematical sciences. These developments show that the stable and reproducible motions that dominated science for centuries do not always symbolize our physical world. Experiments on quite ordinary physicochemical systems at the laboratory scale and the study of simple mathematical models have revealed the existence of instabilities that amplify small effects and drive the system from arbitrarily closed initial states to many possible alternate future states. These nonlinear phenomena are sources of intrinsically generated complex behavior and unpredictability, in the sense that many outcomes of the evolution now become possible. Although these evolutions are governed by a well-defined set of deterministic laws, they exhibit an aperiodic random-looking behavior, called deterministic chaos. Clearly, for such phenomena it becomes meaningless to eliminate the variability and to keep only the mean as being the most representative part of the behavior. Historically, one of the most compelling early arguments demonstrating the presence of chaotic behavior in nonlinear phenomena came from the geosciences. In the now-classic work of meteorologist Edward Lorenz (1963), the existence of deterministic chaos was shown numerically on a simplified model of the phenomenon of thermal convection.

Figure 3.32 depicts a typical scenario of the way the solutions X of a nonlinear dynamical system behave when a parameter λ built into it is varied. At the values λ_1, λ_2, λ_3, . . ., of the parameter, usually referred to as bifurcation points, new branches of solution are generated. These bifurcation cascades often culminate in deterministic chaos. In general, they produce a multiplicity of simultaneously available states, known as attractors. Which of these states in the attractor is actually chosen depends on the initial conditions. This high degree of sensitivity to initial conditions confers a markedly random character on the system, since the initial conditions are history-dependent and may be modified by the fluctuations or by external perturbations. In actual fact, therefore, the dynamics of a multistable nonlinear system

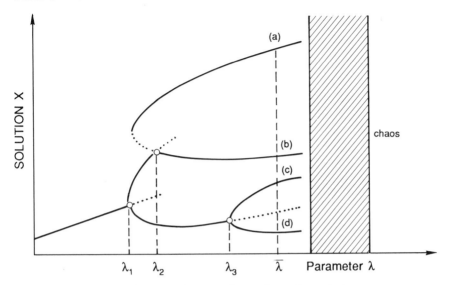

FIGURE 3.32 Typical bifurcation diagram of a nonlinear dynamical system.

exhibit an aperiodic succession of intermittent jumps between coexisting attractors. This view is reminiscent of a great number of natural processes. For instance, it has been used recently to model the principal features of the abrupt climatic changes associated with Quaternary glaciations. As is illustrated below, it should also prove useful in understanding the dynamical aspects of hydrologic time series, on time scales of the order of centuries, such as the abrupt changes in the succession of humid and dry periods in the regime of annual precipitation.

Can the dynamics of multistable nonlinear systems suggest new physical insights into the patterns of annual rainfall time series? Will these insights bring a greater potential for drought prediction?

Let us choose the onset of drought in the western Sahel as a case study (Demaree and Nicolis, 1990). Figure 3.33 depicts a typical time series record. Statistical analysis shows that such a record is nonstationary and that a more or less well-defined transition occurs between plateaus. This transition in the example in Figure 3.33 occurred in the mid-1960s. Actually, an examination of the historical record of the Sahel

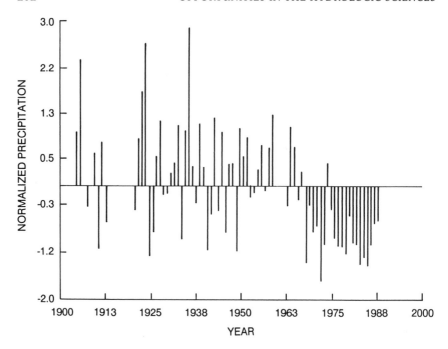

FIGURE 3.33 Normalized rainfall departures, in standard deviations, at the Kaedi station (Mauritania) for the period from 1904 to 1988. SOURCE: Reprinted, by permission, from Demaree and Nicolis (1990). Copyright © 1990 by the Royal Meteorological Society.

region shows that such abrupt changes of rainfall prevailed over several centuries in the past. Following the ideas of the dynamical systems approach explained above, one may wish to stipulate that the annual precipitation time series is a bistable nonlinear dynamical system possessing two stable states, one of quasi-normal rainfall and the other of low rainfall. However, once the system lands in either of these states, it is still subjected to variability arising principally from two sources. The first one includes the imbalances that inevitably exist between such internally generated processes as transport and radiative mechanisms, and the second source arises owing to disturbances of an external origin, such as the sea surface temperature anomalies. These phenomena are perceived by the dynamics as a stochastic forcing, and the evolution of such a stochastic-dynamical system can be expressed in a generic form as

$$\frac{dX}{dt} = f(X, \lambda) + F(t)$$

(the time rate of change of X equals the nonlinear dynamics plus the stochastic forcing).

Here X denotes the variable to be predicted, and f is the nonlinear dynamical part of the evolution, which includes the effects of feedback, of radiation, and so on, and also depends on a parameter λ. The term $F(t)$ represents a stochastic forcing. In the simplest form, $F(t)$ is generally assumed to have no correlations and to have a normal probability law. It is called the Gaussian white noise.

The principal features of the evolution predicted by the above equation may be summarized as follows. Suppose that the system starts in one of its two stable attractors. If the strength of the stochastic forcing $F(t)$ is small, then during a long period of time the system will perform a small-scale jittery motion around a level corresponding to this attractor. But sooner or later, there is bound to be a fluctuation capable of overcoming the "barrier" separating this state from the second available state. When this happens, the system finds itself in another attractor in a very short time interval. Subsequently, it will again undergo a small-scale random motion around this new state until a new fluctuation drives it back to the previous state, or to a third one if such is available. This intermittent evolution looks very much like the record of Figure 3.33. More generally, it provides us with an archetype for understanding other hydrologic processes beyond our specific example, for instance, river flows that seem to exhibit abrupt transitions. Of course, a more quantitative view requires that the function $f(X, \lambda)$ and the noise strength $F(t)$ be known. A minimal model of f should involve nonlinearities giving rise to stable states whose number and characteristics are identical to those of the plateaus found from the statistical analysis of the record. Having chosen the dominant nonlinearity, one can actually determine most of the model parameters from the data.

An interesting question pertains to the residence times, that is, the time the system spends in a given attractor state or, alternatively, to the transition times between attractors. It would obviously be quite interesting to predict such times, since this would be equivalent to predicting the duration of an ongoing drought or to forecasting a forthcoming one. The theory of stochastic processes allows one to make statistical predictions of these times for the systems described by the above equation. Applied to the Sahel record shown in Figure 3.33, this type of an analysis predicts that the mean transition time from the dry state is much larger than the time from the more humid state. The theory also predicts an appreciable dispersion around these mean values.

In conclusion, dynamical systems theory suggests new techniques of data analysis. It also allows us to formulate the key questions pertaining to the dynamical behavior of complex hydrologic phenomena from a novel point of view. New physical insights and predictive capabilities will emerge from such analyses in the future.

Nonlinear Dynamics and Predictability of Hydrologic Phenomena

Weather and climate processes of hydrologic interest, such as rainfall, exhibit a complex and highly variable structure in time and space. The general approach for making predictions of these processes, as, for example, for real-time flash flood forecasting, has been to use nonlinear deterministic equations governing atmospheric dynamics and to solve these equations numerically using high-speed computers. This is called numerical weather forecasting. Within the last two decades the complexity of numerical models has grown commensurately with the capacity and speed of computers. However, despite substantial progress in short-term weather forecasting, the reliability of forecasts has not increased much.

Recent developments in the theory of dynamical systems show that many nonlinear deterministic phenomena are sources of intrinsically generated complex behavior and unpredictability. As explained earlier in this section, solutions of many nonlinear dynamical systems can take any one of many possible states called attractors. Which of these alternate states is chosen by the system depends on the initial conditions. This high degree of sensitivity to initial conditions confers a markedly random-looking character to the evolutions governed by purely deterministic dynamical equations.

Ordinarily, in mathematical modeling or in laboratory experiments, the state (physical) variables are known in advance, and one deals with a well-defined set of evolution laws for these variables. However, this full information is seldom available for a natural system. Rather, only an observed time series of a climatic variable, say, rainfall rate, is available at one or several locations in space. Recent advances in dynamical systems theory have been instrumental in the development of new techniques to provide important qualitative information about process dynamics from the observed time series at one or several locations in space. They do not depend on specific model assumptions and details of the nonlinear dynamics. Therefore an important problem is to learn more about the underlying dynamics of weather and climate processes, independent of any modeling, from the observed time series, and to find to what extent they are predictable.

Are there strange attractors in hydrologic time series? What are the limits of predictability of hydrologic phenomena?

Suppose that $X_k(t)$, $k = 0, 1, \ldots, r - 1$, are the state variables actually taking part in the dynamics. The mathematical space in which these variables take values is called the phase space. X_k's are assumed to satisfy a set of first-order nonlinear equations whose form is unknown, but which, given a set of initial data $X_k(0)$, produce the full details of the evolution. By successive differentiation in time, this set of r equations can be reduced to a single, highly nonlinear, rth order equation for any one of these variables. For example, instead of $X_k(t)$, $k = 0, 1, \ldots, r - 1$, one can take $X_0(t)$ and its $(r - 1)$ successive derivatives to be the r state variables spanning the full phase space. Now, the most important point to notice is that both $X_0(t)$ and its $(r - 1)$ derivatives can be deduced from a single observed time series, $X_0(t_1), X_0(t_2), \ldots, X_0(t_n)$, where t_1 is the initial time, $\Delta = t_2 - t_1 = t_3 - t_2 = \ldots = t_n - t_{n-1}$ is the sampling time, and n is the total number of observations. So, in principle, an observed time series contains sufficient information about the multidimensional character of the system's dynamics.

Important scientific issues, such as the extent of the predictability of a natural system, depend on the nature of the trajectories of the dynamical system in phase space, i.e., the "geometry" of the phase space. In order to identify this geometry from observed time series data, one typically wants an estimate of the minimum number r of variables that captures the essential features of the long-term evolution of the climatic or weather system. This number also denotes the dimension of the phase space. In addition, one wants to test for the possible existence of an attractor in the phase space that represents this evolution.

In a dissipative dynamical system like rainfall, the attractor occupies only a reduced portion of the phase space and therefore has a lower dimension than that of the phase space. One might visualize this scenario with the example of a simple pendulum. Its trajectories lie in a two-dimensional phase space, defined by an angle θ with the vertical direction and the angular velocity $d\theta/dt$. If the pendulum loses energy to friction, then the trajectories gradually spiral inward toward a point that represents the state of no motion. In this case, the attractor is a zero-dimensional point. If the energy supplied to the pendulum exactly balances the energy dissipated by friction, then a steady state is reached in the form of a repeating loop in the phase space. In this

HAROLD EDWIN HURST
(1880-1978)

The Nile is, of course, one of the world's great rivers, and the man sometimes known as "Abu Nil," or the Father of the Nile, shared some attributes with the great river he spent his life studying. Like the Nile, Harold Edwin Hurst had humble beginnings but rose to powerful stature. Like the great river, which slowly shaped the face of the landscape over time, Hurst's most notable contributions resulted from years of patient effort.

Nilologist Hurst was born in Leicester, England, to a family of limited means. Trained mostly in chemistry—and in carpentry by his father, a builder—Hurst left school at age 15. He worked as a pupil-teacher at a school in Leicester and attended evening classes to continue his own education.

Hurst won a scholarship to go to Oxford as a noncollegiate student at age 20, but soon obtained regular admission and began majoring in physics. Such a pursuit must have been difficult for someone with his limited training in mathematics, but one of his professors took special interest in the development of this talented young man. The teacher's faith was rewarded: to everyone's surprise, Hurst won a first-class honors degree and was invited to stay for three years as a lecturer.

In 1906, Hurst left England for a short assignment in Egypt. He remained there for 62 years. He was fascinated with the Nile, and he explored the basin extensively. He believed it was critical for Egypt to design irrigation storage schemes that would carry the area through a series of dry years, and he was one of the first to state the need for a high dam and reservoir at Aswan.

Among specialists, Hurst's name is associated with a statistical method he used to discover a major empirical law concerning long run dependence in geophysics. His location—the Nile basin—contributed to his development of this method. Since he was not pressed by time and had an abundance of data, he was able to compare his data with the standard model of stochastic variability (white noise) through their respective effects on the design of the high dam. Without the aid of computers, this was a tremendous task.

Although the significance of his findings was disputed because no test existed by which the significance could be assessed objectively, finally, at the ages of 71 and 75, he presented two papers on his discovery that firmly established its importance. His work was interesting from both a statistical and mathematical point of view, and his accomplishments as a civil servant far from any major center of learning were inspiring.

case, the attractor is a one-dimensional closed curve in the phase space and is called a limit cycle. Trajectories of many natural systems like rainfall do not converge with time either to a point or to a limit cycle. Even though the attractor has a dimension smaller than that of the phase space, the trajectories do not cross themselves, do not repeat themselves, and contain every possible frequency in a broadband spectrum. To fulfill these conditions, the attractor has to have some strange geometrical attributes. For example, its dimension turns out to be a fraction rather than a positive integer and therefore is known as a fractal. It is called a strange attractor.

The existence of a strange attractor means that trajectories, which are initially close, ultimately diverge into completely different paths. Therefore, beyond this time, predictability is no longer possible. The limits of predictability are set by the rate of divergence of the trajectories from the initial conditions close to one another. This rate of divergence is measured by the so-called Lyapunov exponents. The inverse of the largest positive Lyapunov exponent gives the time limit of predictability. The calculation of these exponents is an area of active research.

Applications of these techniques of phase space reconstruction from time series are beginning to appear in the literature. Some recent examples include the identification of chaotic attractors governing the weather over Western Europe, the climate dynamics of Quaternary glaciations, and the mesoscale dynamics of certain extratropical storms in the United States. These techniques hold the potential to enhance understanding of different dynamic scenarios in diverse hydrologic processes, e.g., river flows, sediment flows, and rainfall, which is necessary both for developing physical descriptions of these processes and for making predictions.

SOURCES AND SUGGESTED READING

Hydrology and the Earth's Crust

Back, W., and R. A. Freeze. 1983. Chemical Hydrogeology. Benchmark Papers in Geology. Vol. 73. Hutchinson Ross, Stroudsburg, Pa., 413 pp.

Back, W., J. S. Rosensheir, and P. R. Seaber. 1988. Hydrogeology. Geology of North America. Geological Society of America, 524 pp.

Bethke, C. M., W. J. Harrison, C. Upson, and S. Altaner. 1988. Supercomputer analysis of sedimentary basins. Science 239:261-267.

Biggar, J. W., and D. R. Nielsen. 1976. Spatial variability of the leaching characteristics of a field soil. Water Resour. Res. 12(1):78-84.

Biot, M. A. 1941. General theory of three-dimensional consolidation. J. Appl. Phys. 12:155-164.

Dagan, G. 1986. Statistical theory of ground water flow and transport: pore to laboratory, laboratory to formation, and formation to regional scale. Water Resour. Res. 22:1305-1345.

de Jong, de Josselin, and S. C. Way. 1972. Dispersion in Fissured Rock. Unpublished report. New Mexico Institute of Mining and Technology, Socorro, N. Mex.

Delhomme, J. P. 1979. Spatial variability and uncertainty in groundwater flow patterns. Water Resour. Res. 15(2):269-280.

Forster, C., and L. Smith. 1989. The influence of groundwater flow on thermal regimes in mountainous terrain: A model study. J. Geophys. Res. 94(B7):9439-9451.

Freeze, R. A. 1975. A stochastic conceptual analysis of one-dimensional ground water flow in nonuniform homogeneous media. Water Resour. Res. 11(5):725-741.

Freeze, R. A., and W. Back. 1983. Physical Hydrogeology. Benchmark Papers in Geology. Vol. 72. Hutchinson Ross, Stroudsburg, Pa., 431 pp.

Freeze, R. A., and J. Cherry. 1979. Groundwater. Prentice-Hall, Englewood Cliffs, N.J., 604 pp.

Gelhar, L. W. 1976. Effects of hydraulic conductivity variations in groundwater flows. In Proceedings Second International IAHR Symposium on Stochastic Hydraulics. International Association of Hydraulic Research, Lund, Sweden.

Hubbert, M. K., and W. W. Rubey. 1959. Role of fluid pressure in the mechanics of overthrust faulting, I, Mechanics of fluid-filled porous solids and its application to overthrust faulting. Geol. Soc. Am. Bull. 70:115-166.

Jury, W. A. 1985. Spatial Variability of Soil Physical Parameters in Solute Migration: A Critical Literature Review. EPRI Report EA-4228. Electric Power Research Institute, Palo Alto, Calif., September.

Jury, W. A., D. Russo, G. Sposito, and H. Elabd. 1987. The spatial variability of water and solute transport properties in unsaturated soil. Hilgardia 55(4):1-32.

Lachenbruch, A. H., and J. H. Sass. 1980. Heat flow and energetics of the San Andreas fault zone. J. Geophys. Res. 85:6185-6207.

Long, C. S. J., P. Gilmour, and P. A. Witherspoon. 1985. A model for steady fluid flow in random three-dimensional networks of disc-shaped fractures. Water Resour. Res. 21(8):1105-1115.

National Research Council. 1984. Groundwater Contamination. Studies in Geophysics. National Academy Press, Washington, D.C., 179 pp.

Norton, D. 1984. Theory of hydrothermal systems. Annu. Rev. Earth Planet. Sci. 12:155-177.

Raleigh, C. B., J. H. Healy, and J. D. Bredehoeft. 1972. Faulting and crustal stress at Rangely, Colorado. American Geophysical Union Monograph Series, Vol. 16, pp. 275-284.

Sibson, R. H. 1973. Interactions between temperature and fluid pressure during earthquake faulting—A mechanism for partial or total stress relief. Nature 243:66-68.

Sun, Ren Jen, ed. 1986. Regional aquifer system analysis program of the U.S. Geological Survey. Summary of Projects, 1978, 1984. USGS Circular 1002, 26 pp.

Theis, C. V. 1935. The relation between the lowering of the piezometric surface and the rate and duration of discharge of a well using ground water storage. Trans. AGU 16:519-524.

Hydrology and Landforms

Ahnert, F. 1976. Brief description of a comprehensive three-dimensional process-response model of landform development. Z. Geomorphol., Suppl. 25:29-49.

Cooke, R. U., and R. W. Reeves. 1976. Arroyos and Environmental Change in the American Southwest. Oxford University Press, London, 213 pp.

Gagliano, S. M., K. J. Meyer-Arendt, and K. M. Wicker. 1981. Land loss in the Mississippi River deltaic plain. Gulf Coast Association of Geological Societies, Trans. 31:295-330.

Horton, R. E. 1945. Erosional development of streams and their drainage basins: hydrophysical approach to quantitative morphology. Geol. Soc. Am. Bull. 56:275-370.

Keown, M. P., E. A. Dardeau, Jr., and E. M. Causey. 1986. Historic trends in the sediment flow regime of the Mississippi River. Water Resour. Res. 22:1555-1564.

Kirkby, M. J. 1976. Tests of the random network model and its application to basin hydrology. Earth Surface Processes 1:197-212.

Kirkby, M. J. 1985. A two-dimensional simulation model for slope and stream evolution. Pp. 203-222 in Hillslope Processes. A. D. Abrahams, ed. Allen & Unwin, London.

Meade, R. H., and R. S. Parker. 1985. Sediment in rivers of the United States. Pp. 49-60 in National Water Summary 1984. U.S. Geological Survey Water Supply Paper 2275.

Potter, P. E. 1978. The origin and significance of big modern rivers. J. Geol. 86:13-33.

Rodriguez-Iturbe, I., and J. B. Valdes. 1979. The geomorphological structure of hydrologic response. Water Resour. Res. 15(6):1435-1444.

Schumm, S. A., and R. W. Lichty. 1963. Channel widening and floodplain construction along Cimarron River in Southwestern Kansas. Pp. 71-88 in U.S. Geological Survey Professional Paper 352-D.

Selby, M. J. 1982. Hillslope Materials and Processes. Oxford University Press, London. 264 pp.

Shreve, R. L. 1966. Statistical law of stream numbers. J. Geol. 74:17-37.

Smith, T. R., and F. P. Bretherton. 1972. Stability and the conservation of mass in drainage basin evolution. Water Resour. Res. 8(6):1506-1529.

Trimble, S. W. 1977. The fallacy of stream equilibrium in contemporary denudation studies. Am. J. Sci. 277:876-887.

Wells, J. T. Subsidence, sea-level rise, and wetland loss in the Lower Mississippi River Delta. In Sea-level Rise and Coastal Subsidence: Problems and Strategies. J. D. Milliman and S. Sabhasri, eds. John Wiley & Sons, New York, in press.

Wells, J. T., and J. M. Coleman. 1987. Wetland loss and the subdelta life cycle. Estuarine, Coastal and Shelf Science 25:111-125.

Willgoose, G. R., R. L. Bras, and I. Rodriguez-Iturbe. 1989. A physically based channel network and catchment evolution model. Report No. 322. Ralph Parsons Laboratory, Department of Civil Engineering, Massachusetts Institute of Technology, 464 pp.

Hydrology and Climatic Processes

Anthes, R. A., J. J. Cahir, A. B. Fraser, and H. A. Panofsky. 1981. The Atmosphere. Third Ed. Charles E. Merrill, Columbus, Ohio, 531 pp.

Bolin, B., B. Doos, J. Jager, and R. Warrick, eds. 1986. The Greenhouse Effect, Climatic Change, and Ecosystems. John Wiley & Sons, New York, 541 pp.

Budyko, M. I. 1974. Climate and Life. Academic Press, New York, 508 pp.

Callendar, G. S. 1938. The artificial production of carbon dioxide and its influence on temperature. Q. J. R. Meteorol. Soc. 64:223-240.

Charney, J. G. 1975. Dynamics of deserts and drought in the Sahel. Q. J. R. Meteorol. Soc. 101:193-202.

Dewey, K. F., and R. Heim. 1981. Satellite Observations of Variations in Northern Hemisphere Snow Cover. NOAA Technical Report NESS 87. U.S. Department of Commerce, Washington, D.C.

Dey, B., and O. S. R. U. Branu Kumar. 1983. Himalayan winter snow cover area and summer monsoon rainfall over India. J. Geophys. Res. 88:5471-5474.

Dozier, R. J., S. R. Schneider, and D. F. McGinnis, Jr. 1981. Effect of grain size and snow pack water equivalence on visible and near-infrared satellite observations of snow. Water Resour. Res. 17(4):1213-1221.

Glantz, M., ed. 1988. Societal Responses to Regional Climatic Change: Forecasting by Analogy. Westview Press, Boulder, Colo., 428 pp.

Hansen, J., G. Russell, D. Rind, P. Stone, A. Lacis, S. Lebedeff, R. Ruedy, and L. Travis. 1983. Efficient three-dimensional global models for climate studies: Models I and II. Mon. Weather Rev. 111:609-662.

Henderson-Sellers, A., and K. McGuffie. 1987. A Climate Modeling Primer. John Wiley & Sons, New York, 217 pp.

Kondratyev, K. Ya. 1969. Radiation in the Atmosphere. Academic Press, New York.

Koster, R. D., P. S. Eagleson, and W. S. Broecker. 1988. Tracer Water Transport and Subgrid Precipitation Variation with Atmospheric General Circulation Models. Report 317. Ralph M. Parsons Laboratory, Department of Civil Engineering, Massachusetts Institute of Technology. March.

L'vovich, M. I., and S. P. Ovtchinnikov. 1964. Physical-Geographical Atlas of the World (in Russian). Academy of Sciences, USSR, and Department of Geodesy and Geography, State Geodetic Commission, Moscow.

Manabe, S., and J. L. Holloway, Jr. 1975. The seasonal variation of the hydrologic cycle as simulated by a global model of the atmosphere. J. Geophys. Res. 80(12):1617-1649.

Meier, M. F. 1990. Greenhouse effect: reduced rise in sea level. Nature 343:115-116.

Namias, J. 1978. Multiple causes of the North American abnormal winter, 1976-1977. Mon. Weather Rev. 106(3):279-295.

National Research Council. 1988. Toward an Understanding of Global Change. National Academy Press, Washington, D.C., 213 pp.

Nicholson, S. E. 1989. African drought: characteristics, causal theories and global teleconnections. Pp. 79-100 in Understanding Climate Change. A. Berger, R. E. Dickinson, and J. W. Kidson, eds. American Geophysical Union, Washington, D.C.

Richardson, L. F. 1965. 1922 Weather Prediction by Numerical Process. Cambridge University Press, New York, 236 pp. (Reprinted by Dover, Mineola, N.Y.)

Schneider, S. H. 1987. Climate modeling. Sci. Am. 256(5):72-80.

Seligman, G. 1936. Snow Structure and Ski Fields. Macmillan, London.

Sellers, W. D. 1965. Physical Climatology. The University of Chicago Press, Chicago, 272 pp.

Street-Perrott, A., M. Beran, and R. Ratcliffe, eds. 1983. Variations in the Global Water Budget. D. Reidel, Dordrecht/Boston, 518 pp.

Wallace J. M., and P. V. Hobbs. 1977. Atmospheric Science: An Introductory Survey. Academic Press, New York.

Walsh, J. E. 1984. Snow cover and atmospheric variability. Am. Sci. 72:50-57.

Warren, S. G. 1982. Optical properties of snow. Rev. Geophys. Space Phys. 20(1):67-89.

Washington, W. M., and C. L. Parkinson. 1986. An Introduction to Three-dimensional Climate Modeling. University Science Books, Mill Valley, Calif. 422 pp.

Hydrology and Weather Processes

Anthes, R. A., J. J. Cahir, A. B. Fraser, and H. A. Panofsky. 1981. The Atmosphere. Third Ed. Charles E. Merrill, Columbus, Ohio, 531 pp.

Bolin, B., B. Doos, J. Jager, and R. Warrick, eds. 1986. The Greenhouse Effect, Climatic Change, and Ecosystems. John Wiley & Sons, New York, 541 pp.

Byers, H. R., and R. R. Braham, Jr. 1949. The Thunderstorm. U.S. Government Printing Office, Washington, D.C., 287 pp.

Chappell, C. F. 1987. Quasi-stationary convective events. Pp. 289-310 in Mesoscale Meteorology and Forecasting. Peter Ray, ed. American Meteorological Society, Boston.

Eagleson, P. S., N. M. Fennessey, W. Qinliang, and I. Rodriguez-Iturbe. 1987. J. Geophys. Res. 92(D8):9661-9678.

Glantz, M., ed. 1988. Societal Responses to Regional Climatic Change: Forecasting by Analogy. Westview Press, Boulder, Colo., 428 pp.

Leary, C. A., and R. A. Houze, Jr. 1979. The structure and evolution of convection in a tropical cloud cluster. J. Atmos. Sci. 36:437-457.

Newton, C. W. 1950. Structure and mechanism of the prefrontal squall line. J. Meteorol. 7:210-222.

Palmer, C. A. 1952. Reviews of modern meteorology—5 Tropical meteorology. Q. J. R. Meteorol. Soc. 78:126-163.

Schneider, S. H. 1987. Climate modeling. Sci. Am. 256(5):72-80.

Washington, W. M., and C. L. Parkinson. 1986. An Introduction to Three-dimensional Climate Modeling. University Science Books, Mill Valley, Calif. 422 pp.

Waymire, E., V. K. Gupta, and I. Rodriguez-Iturbe. 1984. A spectral theory of rainfall intensity at the meso-β scale. Water Resour. Res. 20(10):1453-1465.

Hydrology and Surficial Processes

Bowen, I. S. 1926. The ratio of heat losses by conduction and by evaporation from any water surface. Phys. Rev. 27:779-787.

Brutsaert, W. 1982. Evaporation into the Atmosphere: Theory, History, and Applications. D. Reidel, Dordrecht/Boston, 299 pp.

Buckingham, E. 1907. Studies on the movement of soil moisture. Bureau of Soils Bulletin No. 38. U.S. Department of Agriculture, Washington, D.C., 61 pp.

Dane, J. H., and A. Klute. 1977. Salt effects on the hydraulic properties of a swelling soil. Soil Sci. Soc. Am. J. 41(6):1043-1049.

Dozier, J. 1989. Spectral signature of Alpine snow cover from the Landsat Thematic Mapper. Remote Sensing Environ. 28:9-22.

Dunne, T., and R. D. Black. 1970. An experimental investigation of runoff prediction in permeable soils. Water Resour. Res. 6(2):478-490.

Eagleson, P. S. 1970. Dynamic Hydrology. McGraw-Hill, New York, 462 pp.

Eagleson, P. S., ed. 1982. Land Surface Processes in Atmospheric General Circulation Models. Cambridge University Press, New York, 560 pp.

Harbeck, G. E., Jr. 1962. A practical field technique for measuring reservoir evaporation utilizing mass-transfer theory. Pp. 101-105 in U.S. Geological Survey Professional Paper 272-E.

Hewlett, J. D., and A. R. Hibbert. 1967. Factors affecting the response of small watersheds to precipitation in humid areas. Pp. 275-290 in Forest Hydrology. W. E. Sopper and H. W. Lull, eds. Pergamon, London.

Horton, R. E. 1933. The role of infiltration in the hydrologic cycle. Trans. AGU 14:446-460.

Hursh, C. R., and E. F. Brater. 1941. Separating storm hydrographs from small drainage areas in surface and subsurface flow. Trans. AGU 22:863-871.

Kirkby, M. J., ed. 1978. Hillslope Hydrology. John Wiley & Sons, New York, 389 pp.

Nielsen, D. R., J. W. Biggar, and K. T. Erh. 1973. Spatial variability of field measured soil water properties. Hilgardia 42:215-259.

Pearce, A. J., M. K. Stewart, and M. G. Sklask. 1986. Storm runoff generation in humid headwater catchments. 1. Where does the water come from? Water Resour. Res. 22:1263-1272.

Penman, H. L. 1948. Natural evaporation from open water, bare soil, and grass. Proc. R. Soc. London, Ser. A 139:120-146.

Sutton, O. G. 1953. Micrometeorology. McGraw-Hill, New York, 333 pp.

Sverdrup, H. U. 1937. On the evaporation from the oceans. J. Mar. Res. 1:3-14.

Sverdrup, H. U. 1946. The humidity gradient over the sea surface. J. Meteorol. 3:1-8.

Hydrology and Living Communities

Arris, L. L., and P. S. Eagleson. 1989. A physiological explanation for vegetation ecotones in eastern North America. Report No. 323. Ralph M. Parsons Laboratory, Department of Civil Engineering, Massachusetts Institute of Technology.

Eagleson, P. S. 1982. Ecological optimality in water-limited natural soil-vegetation systems 1. Theory and hypothesis. Water Resour. Res. 18(2):325-340.

Eagleson, P. S., and R. I. Segarra. 1985. Water-limited equilibrium of savanna vegetation systems. Water Resour. Res. 21(10):1483-1493.

Eyre, S. R. 1968. Vegetation and Soils: A World Picture. 2nd Ed. Edward Arnold, London.

Little, E. L. 1971. Atlas of United States Trees. Vol. 1. Conifers and Important Hardwoods. U.S. Department of Agriculture Misc. Pub. 1141. U.S. Government Printing Office, Washington, D.C.

L'vovich, M. I. 1979. World Water Resources and Their Future. (Translated from the Russian). R. L. Nace, ed. American Geophysical Union, Washington, D.C.

Rosenzweig, M. L. 1968. Net primary production of terrestrial communities: prediction from climatological data. Am. Nat. 102:67-74.

Walter, H. 1973. Vegetation of the Earth. Heidelberg Science Library. Vol. 15. Springer-Verlag, Heidelberg.

Whittaker, R. H. 1975. Communities and Ecosystems. Second Ed. Macmillan, New York.

Hydrology and Chemical Processes

Bader, H., R. Haefeli, E. Bucher, J. Neher, O. Eckel, Ch. Thams, and P. Niggli. 1939. Der Schnee und seine Metamorphose (Snow and Its Metamorphism). Translation 14. U.S. Snow, Ice, and Permafrost Research Establishment, Wilmette, Ill.

Berner, E. K., and R. A. Berner. 1987. The Global Water Cycle: Geochemistry and Environment. Prentice-Hall, Englewood Cliffs, N.J.

Environmental Defense Fund (EDF). 1988. Polluted Coastal Waters: The Role of Acid Rain. EDF, Washington, D.C.

Garrels, R. M., and C. L. Christ. 1965. Solutions, Minerals, and Equilibria. Harper & Row, New York, 450 pp.

Garrels, R. M., and F. T. Mackenzie. 1971. Evolution of Sedimentary Rocks. W. W. Norton, New York, 397 pp.

Harriss, R. C., D. I. Sebacher, and F. P. Day. 1982. Methane flux in the Great Dismal Swamp. Nature 297 (5368):673-674.

Jenny, H. 1980. The Soil Resource: Origin and Behavior. Springer-Verlag, New York.

Johannessen, M., and A. Hendriksen. 1978. Chemistry of snowmelt: changes in concentration during melting. Water Resour. Res. 14:615-619.

Kennedy, I. R. 1986. Acid Soil and Acid Rain. John Wiley & Sons, New York.

McCormick, J., and J. Tinker, eds. 1985. Acid Earth: The Global Threat of Acid Pollution. Earthscan, London and Washington.

Sposito, G., ed. 1989a. The Environmental Chemistry of Aluminum. CRC Press, Boca Raton, Fla.

Sposito, G. 1989b. The Chemistry of Soils. Oxford University Press, New York.

University Corporation for Atmospheric Research (UCAR). 1985. Opportunities for Research at the Atmosphere/Biosphere Interface. UCAR, Boulder, Colo.

Hydrology and Applied Mathematics

Demaree, G. R., and C. Nicolis. 1990. Onset of Sahelian drought viewed as a fluctuation induced transition. Q. J. R. Meteorol. Soc. 116:221-238.

Einstein, A. 1926. The cause of the formation of meanders in the courses of rivers and of the so-called Bear's law. Naturwissenschaften, Vol. 14.

Gupta, V. K., and E. Waymire. 1990. Multiscaling properties of spatial rainfall and river flow distributions. J. Geophys. Res. 95(D3):1999-2010.

Horton, R. 1945. Erosional development of streams and their drainage basins: Hydrophysical approach to quantitative morphology. Geol. Soc. Am. Bull. 56:275-370.

Leopold, L. B. 1962. Rivers. Am. Sci. 50(4):511-537.

Leopold, L. B., and W. Langbein. 1962. The concept of entropy in landscape evolution. U.S. Geological Survey Professional Paper 500-A.

Lorenz, E. N. 1963. Deterministic nonperiodic flow. J. Atmos. Sci. 20:131-141.

Nicolis, C., and G. Nicolis, eds. 1987. Irreversible Phenomena and Dynamical Systems Analysis in Geosciences. D. Reidel, Dordrecht.

Rodriguez-Iturbe, I., B. Febres-Power, M. B. Sharifi, and K. P. Georgakakos. 1989. Chaos in rainfall. Water Resour. Res. 25(7):1667-1676.

Shreve, R. L. 1966. Statistical law of stream numbers. J. Geol. 74:17-37.

Steinhaus, H. 1954. Length, shape, and area. Colloq. Math. 3:1-13.

Scientific Issues of Data Collection, Distribution, and Analysis

Hydrologic processes are highly variable in space and time, and this variability exists at all scales, from centimeters to continental scales, from minutes to years. Data collection over such a range of scales is difficult and expensive, and so hydrologic models usually conceptualize processes based on simple, often homogeneous, approximations of nature. Hence a 2,000-km² river basin is commonly modeled as a lumped system that responds as a point with average representative properties. Ground water flow is commonly treated as one-dimensional or two-dimensional. Rainfall is expressed as a mean over large areas, and as depths over periods of a day. Snowmelt runoff volumes are forecast from averages of snow accumulation at a few index plots. These generalized conceptualizations reflect the normal dearth of data, which lack the temporal and spatial resolution to support more detailed modeling. This forced oversimplification is impeding both scientific understanding and management of resources.

In the history of the hydrologic sciences as in other sciences, most of the significant advances have resulted from new measurements, yet today there is a schism between data collectors and analysts. The pioneers of modern hydrology were active observers and measurers, yet now designing and executing data collection programs, as distinct from experiments carried out in a field setting with a specific research question in mind, are too often viewed as mundane or routine. It is therefore difficult for agencies and individuals to be doggedly persistent about the continuity of high-quality hydrologic data sets. In the excitement about glamorous scientific and social issues, the scientific community tends to allow data collection programs to erode.

214

Such programs provide the basis for understanding hydrologic systems and document changes in the regional and global environments. **Modeling and data collection are not independent processes. Ideally, each drives and directs the other. Better models illuminate the type and quantity of data that are required to test hypotheses. Better data, in turn, permit the development of better and more complete models and new hypotheses.** If we accept this synergism, the hydrologic sciences will be well situated for progress, which is needed because recent developments in spatial and temporal models and new data acquisition technology require a rethinking of many of the traditional hydrologic problems. We must, however, reemphasize the value and importance of observational and experimental skills.

To address many of the issues described in Chapter 3, we need new observations of hydrologic phenomena. Some of the current uncertainties in our knowledge of the hydrologic cycle require better understanding of hydrologic processes, but progress in the hydrologic sciences will also depend on improved methods for collecting hydrologic data, more complete and better-organized archives of already-available information, and better mechanisms for distribution and exchange of data, particularly in developing countries and in the international arena. This chapter describes some requirements for and characteristics of hydrologic data, assesses the current hydrologic data base, and then discusses some opportunities to improve hydrologic data and their use.

NEED FOR COLLECTION OF HYDROLOGIC DATA AND SAMPLES

Hydrologic data are needed to measure fluxes and reservoirs in the hydrologic cycle and to monitor hydrologic change over a variety of temporal and spatial scales.

Historically most hydrologic data have been collected to answer water resources questions rather than scientific ones.

Hydrologists use information obtained in laboratories, such as soil particle size, solute concentrations, or electromagnetic spectra, but

HYDROLOGIC IMPLICATIONS OF GLOBAL WARMING

The issue of global climatic warming hangs like a draconian sword over the entire world. Some parameters of the problem are certain, for example, measured increases in the amounts of carbon dioxide and trace gases in the atmosphere. However, the empirical evidence required to confirm that global warming actually is occurring (e.g., the important historic time series of sea surface and global air temperatures) is unfortunately ambiguous and conflicting. Decades of continued careful measurements are needed to isolate the effects of global climatic warming and to identify the location of these changes. This raises interesting questions: Can surrogates be found for use as indirect indicators of global climatic warming? What scientific measurements can hydrologists make that will contribute to the data bases that describe global climatic change?

It is highly probable that an increase in global atmospheric temperatures would cause a rise in the snow line or freezing elevation on the surface of the earth. This fact alone would have a profound impact on the volume and water content of the snow stored in the mountains of the world in general and in the western United States in particular. For example, the California State Department of Water Resources would expect an annual loss of 3 million acre-feet of water stored in snow in the Sierra Nevada. At first glance, a reduction in winter snow storage appears to be a single and unimportant occurrence in the entire hydrologic cycle and in water resources development. However, this reduction in snow storage would affect virtually every aspect of the rainfall-runoff process and of the U.S. western water industry. Currently, hydrologic planners rely on the accumulation of snow during the winter months of December through March, and a subsequent melting of the snow and release of water during the spring months of April through June. The net result of an increase in global temperature would be a decrease in streamflow during the normal snowmelt runoff months of the spring and summer. This change in the monthly timing of available streamflow could dramatically increase winter flooding and significantly decrease the water available during the critical April-through-September season of soil-moisture replenishment via irrigation. The changing streamflow patterns would affect existing surface water reservoirs whose storage capacities were designed to accommodate historic hydrologic runoff patterns. The impact of global climatic change on water resources systems is potentially very serious. The costs of taking countermeasures and repairing damage could have a significant effect on the economies of individual states and of the United States.

The key question for hydrologists now becomes: What immediate steps should be taken in the face of such long-term uncertainty? Certainly one answer lies in a commitment to continue and to expand programs of careful scientific measurements. The resulting hydrologic data bases will enable hydrologists to study and predict the effects of global warming and make appropriate recommendations for social policy.

most of their data must be obtained under field conditions. The reason is that, in addition to the elucidation of microscale processes, hydrologists are concerned with processes that have meaning only at the field scale or over long time scales, such as runoff and sediment yield from drainage basins or continental-scale drought. These requirements make hydrologic data collection complicated and expensive. Despite large financial investments, there remain important questions about hydrologic fluxes and reservoirs that are unlikely to be answered by the incremental growth of instrument networks. Technical and analytical innovations are necessary to overcome the paucity of useful hydrologic data now being collected and collated.

To better characterize the hydrologic cycle requires data in several categories, and the choice of what to measure and where and when to measure influences what hydrologic questions we can investigate.

1. We require information about the fluxes and storage of water in its various phases as it moves through the components of the hydrologic cycle. These include precipitation, snow accumulation and ablation, glacier flow and mass balance, discharge in rivers and streams, movement of ground water, and evapotranspiration. Also needed is information about the transport of solutes and sediments as well as the fluxes of energy that drive the hydrologic processes.

2. Hydrologic data are needed to monitor change, or lack of change, in the quantity and quality of water. The major effect of climatic variability on the humans, plants, and animals that inhabit the earth is felt through the hydrology. Similarly, changes in water chemistry cause concern among users of a water resource and can dramatically affect the fish and other biota that live in lakes and streams (Figure 4.1). Thus we need baseline data, especially in tropical and semiarid areas.

3. In the traditional scientific sense, hydrologic data are needed to test hypotheses and models, and for exploration, to formulate new hypotheses. Hydrologic science can advance as a discipline only if measurement and theory evolve together. Sometimes the mechanisms that govern a complicated hydrologic process are known so poorly that precise data are needed simply to explore the phenomenon; then the next generation of measurements awaits conceptual developments that show which data are essential for testing ideas about how hydrologic phenomena occur. We know only what we measure, and we know what to measure only after some unifying conceptualization of the existing data has pointed the way.

Finally, **the measurement of hydrologic variables is a scientific endeavor itself.** Future progress in hydrologic data collection should result from:

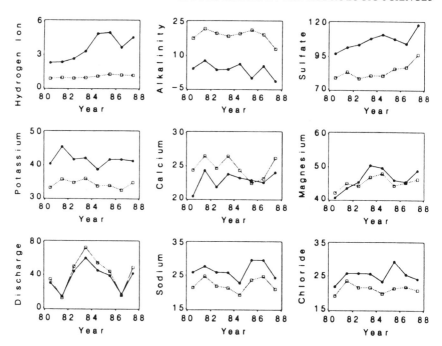

FIGURE 4.1 Changes in ionic concentrations in two streams in Shenandoah National Park, Virginia, 1980 to 1988. Units are microequivalents per liter for concentrations and centimeters per year for discharge. SOURCE: Reprinted, by permission, from Ryan et al. (1989). Copyright © 1989 by the American Geophysical Union.

- coordinated experiments where diverse efforts are pooled;
- technological advances in such fields as remote sensing, instrumentation, and information systems;
- new forms of analysis such as isotope geochemistry, paleohydrology, and improved models of spatial and temporal processes; and
- intensified efforts in design of monitoring networks and examination of data quality and compatibility.

Need to Collect Data at Varying Scales

Hydrologic processes operate over a range of temporal and spatial scales, and important questions exist at time scales from seconds to millennia and space scales from the molecular to the global.

Hydrologic processes operate over a continuum of space and time scales, from those of laboratory experiments to global transport of water and nutrients and from short-lived, transient phenomena to gradual secular variations. Some important questions studied in the laboratory or at scales of a field plot involve interactions between solutes and water or between water vapor, liquid water, and ice. For example:

• The rate of elution of chemical impurities from seasonal snow depends on interactions between solid, liquid, and vapor phases.
• The cycling of hydrogen ions plays a critical role in determining the effects of acidic deposition on wild land and agricultural ecosystems. The largest components of hydrogen ion cycling are consumption by mineral weathering and production by plant roots. These components are difficult to estimate at field scales because we do not know enough about the kinetics of mineral dissolution reactions and biological release processes at these scales. Errors in estimating annual hydrogen ion consumption and production rates can be as large as the estimated annual input rate from acidic deposition.

At the same time, **our current knowledge of major fluxes of water in the hydrologic cycle involves large uncertainty.** For example:

• The mean annual discharge of the Amazon River is about 200,000 m^3/s. Typical error estimates for the measurements are 8 to 12 percent, i.e., 16,000 to 24,000 m^3/s, a rate slightly higher than the mean annual discharge of the Mississippi River.
• Data show that sea level is rising slightly, but our investigations into the source of this rise, and the accuracies of our predictions of the future, are hampered because our measurements of the snowfall and iceberg calving from the Antarctic ice sheets do not tell us definitely whether the Antarctic ice volume is growing or shrinking. Thus the proportion of the water attributed to each of the sources causing this rise in sea level is not confirmed. How much comes from Antarctica and Greenland, from thermal expansion of the ocean waters, from shrinking alpine glaciers, and from depleted ground water reservoirs?

Data are needed at a variety of scales, and the spatial and temporal scales of available data restrict the questions that can be investigated. As is described in detail later in this chapter, the information is better for some hydrologic fluxes and reservoirs than for others. For most fluxes, however, a fundamental block to progress is our poor knowledge of how to interpolate between measurement points and how to extrapolate from few data points. For example:

220 OPPORTUNITIES IN THE HYDROLOGIC SCIENCES

• Depths and water equivalences of a snowpack are measured at many snow courses in cold regions, but it is possible to use these data only as crude estimates of the water content of a regional or basin-wide snowpack.

• Topographic influences on rainfall, evaporation, and soil moisture are poorly documented at scales varying from individual hillslopes to entire mountain ranges.

An additional important issue in the sampling of hydrologic processes is the structure of the statistical fluctuation that the processes have at different scales of measurement. How do the mean and variance of annual rainfall change as a function of the area over which the estimation takes place? How do the mean and variance of evapotranspiration depend on the time scale considered? What is the combined effect of time and space scales on the statistical properties of hydrologic variables?

There is an urgent need to

1. quantitatively characterize the fluctuations of hydrologic variables at different time and space scales; and
2. design data collection programs that will allow the study of theoretical constructs, described in Chapter 3, to structurally link the fluctuations at different scales.

Need to Develop Accurate Hydrologic Data Bases to Improve Scientific Understanding

Detection of hydrologic change requires a committed, international, long-term effort and requires also that the data meet rigorous standards for accuracy.

Synergism between models and data is necessary to design effective data collection efforts to answer scientific questions.

Development of scientific theory in the absence of supporting facts does not lead to understanding and can result only in conjecture. The primary sources of facts for the hydrologic sciences are the measurements that are made of hydrologic and ancillary variables. Most hydrologic data are collected by government agencies for a variety of purposes—only one of which is the development of hydrologic understanding. Historically, the major providers of funding for hydrologic data collection have expected that the resulting data would be useful in setting water policies, developing water resources plans, designing water resources systems, operating the structures that make up such systems, and monitoring the management of water resources. Increased hydrologic understanding can and does contribute to improved information for these utilitarian purposes, but the design of hydrologic data networks seldom has as a primary objective the betterment of basic hydrologic understanding. Therefore, the data needs of the hydrologic scientist almost certainly will not be fully satisfied by the existing data networks that are supported primarily for operational and accounting purposes.

The existing data networks should be viewed by hydrologic scientists as opportunities upon which they can build. To optimize these opportunities, it is first necessary to define the characteristics of the data sets that hydrologic scientists need. These characteristics include the variables to be measured and the locations, frequencies, durations, and accuracies of the measurements. They should be derived from knowledge about the hydrologic phenomena to be explored and from the hypotheses to be tested. Allocation of the resources available for data collection must seek complementarity between the scientific and operational data sets. However, the operational networks often change in character because of changing operational demands for data or because of budgetary pressures on the financial sponsors of the data networks. These changes most often are manifested as discontinuities in the time series of measurements, as shifts in the location of the measuring site, or as changes in the accuracy of the data. Thus a full measure of complementarity is an illusive objective but a worthy one that can be approached by adequate communication between research scientists and managers of data collection programs.

Important hydrologic changes may be subtle or may be difficult to detect because of large interannual or inter-event variation, and spatial and temporal scales of available data restrict the questions that can be investigated. Some important processes are transient—short-lived but repeated. Fluxes and reservoirs of water, energy, solutes, and sediment are monitored most intensively over those parts of the world that are humid-temperate, densely populated, and industrialized, but measurement networks are particularly sparse over the oceans and in regions that are subhumid, tropical, at high elevation, or lightly populated.

Need to Collect Long-Term Hydrologic Data

Long-term monitoring and the use of paleohydrologic records are fundamental to understanding the role of extreme events in hydrologic systems.

The need for long-term measurements is becoming clearer in our investigations of environmental change, including hydrologic change. Some disciplines, such as paleontology and historical geology, have depended for their existence on the availability of data spanning long periods, from 100 million to 2 billion years. Other disciplines that have traditionally focused research over shorter time frames, such as the environmental sciences, now stress the critical importance of long-term records.

Despite the increasingly recognized importance of data records of long duration, only a few dedicated research organizations have successfully maintained high-quality data collection efforts over periods of 50 to 200 years. Furthermore, these organizations have experienced difficulty in committing limited research monies year after year to an activity that is frequently termed "monitoring," often with pejorative overtones.

Dams have been built in many areas of the world and the water has been allocated for power generation or irrigation based on only a few years of data, with the too frequent result that the anticipated volumes of water have been available only in years of above-normal runoff. But many scientific questions justify the collection of long-term data. Two general areas for which long-term hydrologic data are specifically needed are discussed briefly in the remainder of this section, but these examples are not meant to be exclusive or exhaustive.

Understanding Hydrologic Behavior and Hydrologic Change

Long-term data are required to understand the basic hydrologic behavior of natural landscape units. In most humid areas, we do not understand well enough the relationships between rainfall, evapotranspiration, streamflow, and long-lived vegetation such as forests. Research efforts have typically focused on only a short segment during the life span of forest stands that may exceed a century. Moreover, in areas of low rainfall, where the occurrence of rain exhibits high variability over time and space, understanding of such fundamental relationships is even less complete because sufficiently long data records are not yet available to separate the inherent spatial and temporal variability of the processes involved. We do not fully understand, for example, how evaporation and soil moisture are regulated in such situations. The need for long-term data is particularly acute for analyses that focus on hydrologic behavior at the continental spatial scales and on long time scales.

Detection of hydrologic change requires long-term data sets of greater quality and reliability than are normally needed in the investigations of processes. When we measure rainfall for such purposes as flood forecasting, modest changes in the techniques, such as movement or redesign of the gage, do not affect the usefulness of the data for telling us whether to expect a flood on the river. However, when we try to use the same data to identify a long-term trend that is superimposed on the natural year-to-year variability, movement or redesign of the gage may introduce artifacts into the data set, and these may be falsely identified as trends or may disguise hydrologic change.

Identifying Extreme Events

Identification and analysis of hydrologic extremes, such as floods or droughts, are needed to understand the functioning of human societies as well as natural and managed ecosystems. In most hydrologic processes the extreme events often have the greatest effects on both systems and humans. Because they are infrequent in occurrence, they are poorly represented in all but the longest hydrologic records; only a few data sets contain enough extreme events to allow a precise estimation of their return periods. Moreover, the dynamics of extreme events are hard to measure; stage versus discharge relationships for gaging stations are usually not calibrated for high stages, and scouring of the channel during such flows makes extrapolation of rating curves for lower stages prone to error.

Flood frequencies and drought recurrences may be well defined for mid-range events, but the tails of the distributions are poorly quantified, both in temporal distribution and magnitude. A series of extreme events may represent just that, a combination of unlikely probabilities, or it can show a change in climate, whereby the events are no longer extreme but merely normal events within a new population.

A good example is provided by analysis of the 1985-1986 drought in the southeastern United States. Estimation of the severity and interval of likely recurrence for this drought was made possible by

WATER RESOURCES MANAGEMENT

Hundreds of billions of dollars have been invested since 1890 in federal, state, local, and private water projects. Dams, hydroelectric power stations, levees, channel improvements, navigation locks, and water treatment plants are examples. Now the best reservoir locations have been used, we are more sensitive to environmental issues, and the annual investment in new projects is only a small fraction of that in past years. Attention has shifted toward nonstructural approaches to managing our water resources.

Today an important challenge faces this nation. The challenge is to improve how this vast investment is operated, taking into account economic, environmental, and social values. Increases in annual benefits of only a few percent would be worth tens of billions of dollars and could preserve environmental values that were ignored decades ago. Historically, water projects have been planned to serve one set of purposes but operated to serve another. One example is the Tennessee Valley Project. Originally developed for flood control, it now produces hydroelectric power with benefits far greater than those resulting from flood control. This difference between what is planned and what is actually done suggests a management opportunity, especially since much of the planning occurred decades ago.

A major federal water resources agency, the U.S. Bureau of Reclamation, recognized this challenge and formally realigned its mission away from construction and toward improved management and operation of its vast prior investment in the 17 western states. The nation's largest water resources agency, the U.S. Army Corps of Engineers, has not taken such a formal step, but many of its district and division offices are finding new nonstructural solutions. Management conflicts arise among competing uses for water such as irrigation, municipal supplies, hydropower, power cooling, wildlife preservation, fisheries, water quality control, industrial use, navigation, and recreation. Optimal management of the nation's water resources is the key to resolving these conflicts; better information is the key to improved water management.

The importance of water resources management was highlighted by the 1988 drought. Print and electronic media brought us clear images of the consequences of drought—barges stranded on sandbars in the Mississippi River, empty reservoirs, and withered corn fields. The drought illustrated dramatically that the availability of water involves risk and that the risk seems to be increasing. How will increases in atmospheric carbon dioxide and other greenhouse gases change the climate? What effects will such changes have on our water resources?

Some of the information needed to improve water resources management will come from improved application of existing technology. For example, new data systems using automated surface observations, satellite communications, weather radar, and satellite imagery are beginning to be used. Advances in computer technology, especially computer graphics and geographic information systems and data bases, are bringing new opportunities to make better decisions.

But the important information required for improved water resources management must also come from improved hydrologic science. Data systems are costly and can be justified only when there is an adequate science base to assure that the information will be used well. Major federal data programs, such as the stream gaging program of the U.S. Geological Survey, are continually under severe budget pressures. Moreover, it is never possible to measure all that needs to be known. The only alternative is to develop scientific methods to infer what needs to be known from what can be measured.

the availability of high-quality hydrometeorological records maintained continuously for a site since 1934. An even longer precipitation record, 110 years, was located for a nearby station. Whereas the 1985-1986 drought was the most severe in the 53-year record, the 110-year record revealed five periods of even less rainfall before 1934. This information substantially altered the interpretation and implications of the 1985-1986 drought, showing it to be a much more common event than it was first considered.

Need to Collect Data Worldwide to Address Global Hydrologic Issues

Useful hydrologic data representing processes at the global scale are sparse.

Most hydrologic data have been collected with local-scale (or at best national-scale) questions in mind. Although these questions remain important, hydrologists have begun to recognize subtle hydrologic effects that have important consequences for human affairs. The uneven distribution of hydrologic monitoring stations around the world makes it difficult to study the simultaneity of trends and the full extent of widespread changes that are suspected to be linked. It is therefore necessary for international scientific agencies to assess the distribution and quality of hydrologic data collection around the world. For those hydrologic variables that translate into recognizable electromagnetic features that can be viewed from aircraft or satellites, remote sensing methods must be used to fill in the vast areas between ground-based stations. Because it is not possible with foreseeable technology and resources to monitor hydrology over the entire earth, some important decisions must be made to deploy instruments that will help in answering hydrologic questions of global significance.

Data on water quantity—flows and storages—are usually collected in attempts to define total or partial water budgets for drainage basins. As the size and complexity of drainage basins increase downstream, it becomes increasingly difficult to collate the various data sets— often gathered by different agencies or even different nations—into a useful water budget with a clear definition of the various processes of water transfer and their interactions. It is more difficult to reconstruct the influence of, for example, land use changes on the flow and sediment transport of the Mississippi River than it would be to carry out the same task in a small, forested drainage basin in the southern Appalachians.

Spatial and Temporal Issues in Hydrologic Problems

A fundamental block to progress in using most hydrologic data is our poor knowledge of how to interpolate between measurement points.

Some important hydrologic changes may be difficult to detect because of large interannual or inter-event variation; others are difficult to measure because they involve rare, short-lived, catastrophic processes.

A fundamental block to progress in using most hydrologic data is our poor knowledge of how to interpolate between measurement points. For example, depths and water equivalences of a snowpack are measured at many snow courses in cold regions, but it is only possible to use these data as crude estimates of the water content of a regional or basin-wide snowpack. Strong spatial gradients caused by topography or vegetation on snowfall, interception, redistribution, and melt are not explicitly modeled in forecasting water supply, runoff rates, and soil-moisture recharge, and in mountainous basins rain gages are often poorly distributed in the higher elevations (Figure 4.2). Topo-

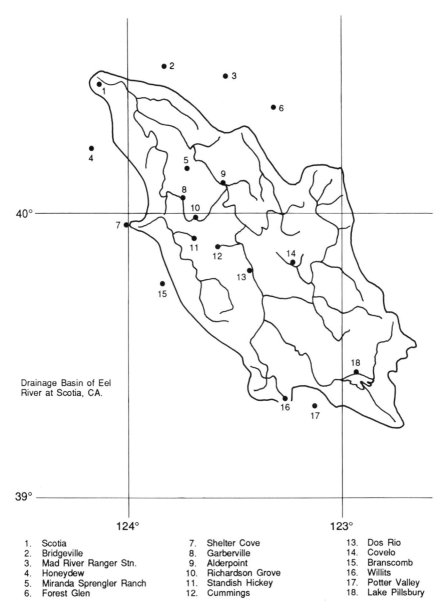

FIGURE 4.2 Rain gages in the Eel River basin at Scotia, California. (At first glance the rain gages appear to be well distributed, but most are in the river valleys, and so precipitation is not measured in the higher elevations.)

graphic influences on rainfall, evaporation, and soil moisture are poorly documented at scales varying from individual hillslopes to entire mountain ranges, and there is currently little understanding of the relationship between regional, time-averaged rainfall rates and the frequency and magnitude of storms. Understanding how to interpolate from a few measurement points requires some thorough field studies of the processes responsible for spatial variability and of the relationship between long-period averages and hydrologically significant events.

Large-scale hydrologic processes such as the coupling between global atmospheric circulation and the North American seasonal snowpack or regional droughts are not well understood or predictable. Understanding them requires simultaneous measurements of phenomena that have traditionally been within the purview of different disciplines or agencies. There is a need to undertake simultaneous large-scale measurement programs within the context of specific hypotheses about these couplings. Careful design and international cooperation will be necessary to develop such programs in a useful way.

Important hydrologic changes may be subtle and difficult to detect because of large interannual or inter-event variation. Recognition of such changes or their absence therefore requires carefully designed, long-term monitoring networks. The paucity of suitably long instrumental records requires the use of historical stratigraphic, botanical, and geochemical records of hydrologic change. The development and quantitative interpretation of these noninstrumental records are in their infancy, but they promise much useful insight into the frequency and magnitude of climatic and hydrologic fluctuations, floods, and sediment yields.

Some hydrologic events involve rare, short-lived, catastrophic processes, such as the influence of the eruption of Mount St. Helens on flooding and sedimentation along the Columbia and Cowlitz river valleys (Figure 4.3), or the effects of intense forest fires on runoff and erosion during succeeding wet seasons, or the release of toxic chemicals after industrial accidents. The probability is low that an instrumental network will be in place to record such an event adequately. Therefore, government agencies and granting agencies need to be able rapidly to mount coordinated field studies to collect data when such transient processes occur.

Collection and Archiving of Selected Water Samples

The archiving of selected water samples will allow better analysis of chemical and biological changes in aquatic habitats.

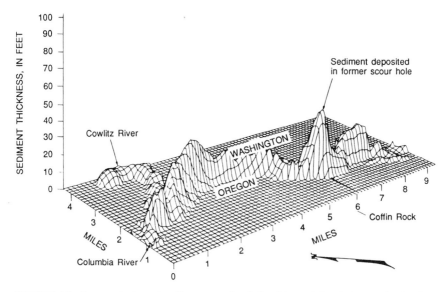

FIGURE 4.3 Accumulation of sediment on the Columbia and Cowlitz river beds after the eruption of Mount St. Helens on May 18, 1980. SOURCE: Reprinted from Haeni (1983) courtesy of the U.S. Geological Survey.

Hydrologic data needs involve more than requirements for numerical information. The archiving and rapid retrieval of samples for future analyses will be needed for the continuing analysis of environmental conditions in aquatic habitats. For example, an essential feature recommended for the National Science Foundation's Long-Term Ecological Research (LTER) Program is the archiving of chemical and biological samples. Established museums have served historically as repositories for biological materials. Unfortunately, such collections, although usually catalogued and curated well, have been disproportionately directed toward terrestrial organisms, or toward only the terrestrial life stages of aquatic species. Thus special initiatives will be necessary to ensure that appropriate aquatic biological materials are archived and curated, and cryogenically or chemically stabilized water samples from carefully selected sites should be archived for future analysis. If such samples were available today from past decades, they could be used to evaluate changes of chemical parameters in aquatic environments.

STATUS OF HYDROLOGIC DATA

Knowledge of the distributions in space and time of water, solutes, and sediments is needed for investigation of scientific questions, management of water resources, and environmental decision making. Industrial and agricultural development in temperate regions has led to intensive measurement of hydrologic variables, and environmental managers have been able to manage resources in such regions, with large errors occurring only occasionally. Many developed nations have extensive, governmentally sponsored programs to monitor the quantity and quality of water resources. Measurements include monitoring of ground water, reservoirs, streamflow, precipitation, and evapotranspiration, and, to a lesser degree, the amount of dissolved and particulate contents in the water.

Long-term records are rare, seldom exceeding 50 to 80 years in length, and often the sites at which data are collected have been moved, or the hydrology anthropogenically altered, so that records are not homogeneous. Records from the few long-term stations that have collected data over several centuries and from paleohydrology show substantial changes in hydrologic regimes. Our knowledge of the frequencies of extreme events is usually uncertain.

Availability of Hydrologic Measurements

Hydrologic data networks are best developed in the humid-temperate, mid-latitude, industrialized nations.

Detection of hydrologic change requires accurate, long-term measurements, and reliable intercomparisons of measurements made in different areas or many years apart are crucial.

Although most of the developed regions are classified as having a humid-temperate climate, the demands from the population, agriculture, and industry for water of appropriate quality are not always met. In semiarid areas, such as the American Southwest and some metropolitan regions, projections indicate that demand will exceed supply within decades, requiring either conservation measures or more importing of water. In less developed, often tropical regions, data and knowledge of local hydrologic cycles are woefully lacking. The tragic realities of drought prove that precipitation and water supplies are inadequate, but there are few data that could be used to show potential carrying capacity even if judicious resource management were to be undertaken. In wetter tropical regions, major land use changes are under way that may alter the hydrologic cycle, and there is little quantitative information to show the natural variation on which the changes will be overlaid.

Precipitation and streamflow are two of the most fundamental quantities and are the two most frequently measured. Sufficient data are available in developed regions to estimate mean values for precipitation and surface runoff. However, only rarely are precipitation data comprehensive enough to quantify spatial distributions that are crucial for some applications, such as precise flood forecasting. This is especially true for regions where floods may be produced by unevenly distributed convective storms. Normal ranges of streamflows are often known for particular drainage basins, but high floods and extremely low flows are measured much less precisely.

As humans use more exotic chemicals in industry and everyday activities, the opportunities for these chemicals to enter the hydrologic regime increase greatly. Complete analysis of water for all common and exotic solutes is expensive, but for water used for human consumption and industrial processes, such analysis is becoming more necessary. Contamination of surface water and, even more so, ground water can render resources permanently unusable. More data on solutes are urgently needed to document contamination and to lead to the development of better preventive measures.

Even in developed, more intensely measured regions the data are seldom adequate for reliable forecasting and analysis of water quality. The problems are of a dual nature: the supply of water and its quality are one problem; current and projected use is also uncertain. Recognized contamination and loss of resources are increasing at the same time that water use is increasing. Sometimes, data are available to show the probabilities for long-term availability of the water supply and the variation through time. Usually, however, the record is too short to show the range of likely natural variation or too limited spatially for comprehensive hydrologic planning.

Fluxes and Reservoirs of Water, Solutes, and Sediment

Fluxes and reservoirs of water are measured routinely, but knowledge of spatial and temporal distribution is not adequate everywhere.

HYDROLOGIC SAMPLING IN PRISTINE WATERSHEDS

Beginning in the late 1950s the U.S. Geological Survey sought to establish a nationwide network of stream gages and water-quality-sampling stations in basins that remained near their natural state. Several prominent hydrologists of the time were concerned that hydrologic data were being collected almost exclusively for immediate applications (toward such practical needs as flood control and irrigation) and that, consequently, the existing hydrologic data base pertained mostly to rivers and streams affected by industrial and agricultural development. Their rationale for creating a network of hydrologic bench marks was that quantitative and qualitative changes in rivers stemming from climatic and other natural factors are difficult to distinguish from changes brought about by human activity. Moreover, they predicted that the need to make this distinction would increase with time. By the late 1960s, some 50 sampling stations in largely undeveloped basins had been identified to provide data that could be compared to those collected at the several thousand other stations under regular operation. During the 1970s, as interest in measuring environmental degradation increased, the foresight displayed in these early steps to define natural hydrologic conditions became apparent.

Several problems have been encountered, however, both in developing the hydrologic bench mark network and in using the data collected for their intended purpose, i.e., for establishing hydrologic baselines. It became clear early in the program that the original goal of finding 100 stations in pristine basins of variable size and setting was unrealistic. For most of the history of the network, the number of sampling stations has remained at about 50, located mostly in basins of 25 to 250 km² in area. Larger basins unaffected by development were difficult to find; in smaller basins, natural hydrologic conditions were considered too ephemeral to provide stable records. Even among the 50 basins that were found, compromises had to be made in the original selection criteria, including accommodation of moderate levels of human habitation, agriculture, and logging.

The narrow range of basin sizes represented in the network leads to problems in making comparative use of the data once they are collected. Because stream quantity and quality tend to vary greatly with basin size and with geological and climatological history, direct comparisons cannot generally be made between the streamflows and chemical concentrations measured at these stations and those observed in basins outside the size range. Thus, while the program has been successful in collecting data from minimally developed basins in a wide range of physiographic, geochemical, and ecological settings, the more limited range of basin sizes represented in the network remains a serious problem in using the data to define baseline conditions for streams and rivers in general.

Perhaps the ultimate limit on the ability of hydrologists to find streams with pristine water quality stems from the ubiquitous effects of airborne pollutants on the chemistry of streams. Industrial and agricultural emissions of a variety of materials routinely travel hundreds of kilometers in the atmosphere before dry or wet deposition to the surface, and significant amounts of this matter are ultimately carried to streams by runoff and ground water. Knowledge of the long-range effects of air quality on water quality has increased as a result of research conducted over the past decade on the effects of acid rain and snow on streams and lakes. It is not surprising, therefore, that some of the most successful applications of data from small, undeveloped watersheds have been in measuring the effects of sulfur and nitrate emissions on stream chemistry. The value of these studies has been enhanced by the availability of 20 years of continuous data on stream chemistry to correlate with the gradual changes that have occurred in industrial and agricultural emissions (Figure 4.4 illustrates the case for sulfur).

In summary, many hydrologists have concluded that sampling in undeveloped basins does not provide a distinct set of hydrologic data that can be used alone to define pristine conditions. Instead, these data represent one extreme in the wide range of effects that human activity has had on hydrologic conditions.

Knowledge of the fluxes of solutes and sediment is usually poor.

Routine measurements of soil moisture are sparse.

Rainfall

Rainfall is routinely measured throughout the world, but obtaining solid knowledge of its spatial and temporal distribution is hampered by a diversity of observing standards and an erratic pattern of observing networks. Although in some parts of the world rain gages have operated for over two millenia, extensive coverage exists for one to two centuries at best. As with other hydrologic data, coverage is poorest in arid, semiarid, tropical, and highland regions and over the oceans.

Like many other hydrologic parameters, rainfall is highly variable in time and space. Thus large-scale fields are difficult to derive from

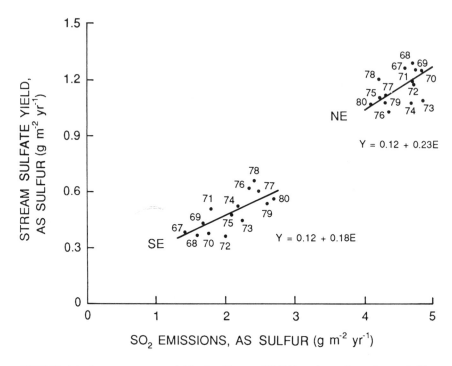

FIGURE 4.4 Average stream yield of sulfate at USGS bench mark stations. Sulfate data are plotted against regional (Northeast, Southeast) sulfur dioxide emissions for the years 1967 to 1980 (two-digit numbers). During the period, both emissions and stream yields of sulfur decreased with time in the Northeast but increased in the Southeast. Both regressions are significant at the 0.01 level. SOURCE: Reprinted, by permission, from Smith and Alexander (1986). Copyright © 1986 by Macmillan Magazines Limited.

point gage measurements, many of which are made at unrepresentative sites. Moreover, data on time scales of less than a month are difficult to obtain, although for many studies and applications, data recorded daily or even hourly are necessary. Gage data are also influenced by the characteristics, location, and exposure of the instrument.

Gradually, satellite methods to estimate rainfall are being developed to rectify these problems, but so far all are highly empirical and cannot readily be applied to locations other than those where the metals have been tested and calibrated. None provides a high degree of accuracy. Immediately needed improvements are the development of methods to optimally combine satellite, radar, and ground measurements with statistical theory to produce large-scale rainfall fields, and the creation of better archives of already existing data.

JAMES CHURCH
(1869-1959)

In 1892 James Church began teaching at the University of Nevada at Reno as a professor of Greek and Latin. He was also an avid outdoorsman who undertook many daredevil, wintertime ascents in the Sierra Nevada. What started out as pure adventure quickly became a passion of some considerable significance in the world of hydrology.

Church's love of the snowy mountains soon led to his spending a great deal of time in wintry landscapes. Aware of his reputation, a local power company asked him to make measurements of the snow deposits so that forecasts of the spring runoff could be made for their power-generating facilities. Church quickly recognized that measurements of snow depth alone were not enough because of snow's variable density, and he developed the Mt. Rose snow-sampling tube. If core samples were extracted from the snow cover, the snow water equivalent could be determined to give the melted equivalent of the snow deposit.

In response to the important need for accurate forecasts of snowmelt runoff, Church gave up his academic position to establish a set of snow survey courses in Nevada. Subsequently, he was involved with the initiation of surveys in other states before exporting his approach to countries around the world. The greatest challenge came in 1947 when he went to India to apply his methods to the world's highest mountains, whose snow fields feed some of the world's greatest rivers.

Church was no mere technician. Along the route from professor of classics to snow scientist, he was sufficiently well respected to be elected president of the International Commission of Snow and Ice, where he had some influence over the rapidly developing field of glaciology. While his methods may seem primitive in this day of remote sensing and easy communications, Mt. Rose samplers and their many derivatives are still one of the fundamental tools of snow and ice data collection today.

Snow Accumulation and Ablation

Snow accumulates as a seasonal cover before it melts to produce water for runoff and ground water recharge. Data on snow cover are collected by a variety of organizations in different countries for different purposes with little exchange of information. In the past, measurements were made at selected points, but remote sensing may allow large-scale measurements if the technological problems of interpreting microwave signals can be solved. This would allow data collection in remote and difficult environments such as mountainous or arctic regimes.

Few stations around the world collect the data needed to analyze the energy balance of a melting snow cover, and many of those that do are experimental and short term. Critical information about properties like albedo is rarely available, and information necessary to drive streamflow-forecasting models is not available except locally. Recent interest in atmospheric deposition and the chemistry of snow cover has stimulated collection of data in some regions, but no consistent data sets are available on regional or global scales.

Surface Runoff

Surface runoff is measured throughout the world. However, the spatial and temporal distributions of these data are erratic and strongly related to the local level of development. In many developing countries, surface-runoff measurements are spotty in both space and time, whereas most developed countries have reasonably dense data networks that have existed for several decades. In the United States there are more than 10,000 locations at which daily records of surface runoff have been computed and archived for a decade or longer.

In addition to the spatial and temporal diversity of surface-runoff data, the accuracy of the data is also variable. Generally, the accuracy of surface-runoff data is better for mid-range flow rates than it is for either high flows (floods) or low flows. Data accuracy also generally deteriorates in extreme environments, such as arid alluvial streams or ice-affected streams in alpine, subarctic, or arctic areas.

Soil Moisture

The temporary subsurface storage of water in the vadose zone usually coincides with the rooting zone of plants. From the soil, the moisture will either be returned to the atmosphere, temporarily stored in vegetation, or percolated to the saturated zone. Information about soil moisture and its spatial distribution would be of great value in delineating the dynamic nature of hydrologic processes, but except in experimental basins, soil moisture is never measured routinely, and data on its spatial variation are extremely scarce.

Soil moisture is hard to measure. Gravimetric measurements in the field are time-consuming, and soil moisture varies markedly over scales of a few meters. Progress in the development of better methods for measuring soil moisture in the top meter of the soil, in areas of both woody and grassland vegetation, is desperately needed. Remote sensing in the microwave wavelengths offers some hope, but the effects of soil, vegetative, and topographic properties need to be worked out.

Subsurface Water Below the Vadose Zone

The basic variable measured in ground water surveys is the water level in observation wells. From these data it is possible to estimate ground water fluxes if information on aquifer properties is available. Various levels of sophistication of analysis are possible, ranging from a simple application of Darcy's law, through flow nets, to use of computer simulation models. It is not possible to estimate fluxes or storage volumes without complementary geological data on properties of the medium. With a sparse data base, there can be considerable uncertainty in these estimates. Rates of ground water recharge or discharge are rarely monitored, although discharge measurements from ground water springs are sometimes made.

Most countries maintain regional networks of monitoring wells in principal aquifers used for water supplies. Summaries of well yields and total withdrawals are also available. Although some hydrographic records extend back over 100 years, most data postdate the mid-1940s. Data density decreases markedly for deeper aquifers. In regions of sparse population or in areas without significant potential for ground water supplies, background data are scarce.

On a continental scale, no inventory exists to define the extent and quality of the total volume of subsurface water in storage. Similarly, while fluxes through the ground water reservoir have been characterized for some major aquifer systems (Figure 4.5), no detailed assessment of fluxes is available on a continental scale.

Evapotranspiration

In spite of its importance in the hydrologic cycle, there are no routine measurements of evapotranspiration. Although the necessary technology has been developed and is available, the instruments still require special expertise to operate and would be costly to deploy in a network of stations for long-term global operation. There is some hope that the anticipated developments in electronics and remote sensing from satellites will bring progress in this area. While a solid data base is lacking, useful information on evapotranspiration can be derived from published maps depicting its approximate long-term average distribution, together with other components of the water balance in different parts of the world.

At a larger landscape scale, evapotranspiration estimates based on remotely sensed data in the thermal-infrared region of the electromagnetic spectrum may be possible. There is a need to examine the relationship between thermal-infrared radiation emitted from a plant canopy and

FIGURE 4.5 High Plains aquifer. Contours show the water table elevation for the aquifer, with ground water flow from the west toward the east. SOURCE: Reprinted, by permission, from Skinner and Porter (1989), Figure 10.9. Copyright © 1989 by John Wiley & Sons, Inc.

its temperature structure. If remotely sensed temperatures of forest canopies were combined with existing atmospheric models, then evapotranspiration might be predicted for broad units of the landscape. Methods could be validated for accuracy and temporal resolution through comparisons with site-specific watershed studies where hydrologic balances are precisely documented. Much of the work in the large-scale, coordinated field experiments—such as HAPEX, FIFE, and GEWEX—addresses this question of the measurement of evapotranspiration over the spatial scales of large drainage basins.

Fluxes of Energy

In spite of the fundamental influence that the processes of surface energy exchange have on the hydrology and climate of the earth, data required to estimate the energy fluxes from the land surface are usually available only for areas where intensive field experiments are under way.

Knowledge of the spatial distribution of energy exchange at the land surface is important if we are to estimate many of the fluxes of interest in the hydrologic cycle. Evaporative fluxes of water in particular are driven by fluxes of energy, which are sometimes easier to measure. Therefore we need information on the exchange of solar and thermal-infrared radiation, latent and sensible heat, and heat flux into or out from the soil or snow. A major influence on the atmospheric circulation and the distribution of precipitation is the distribution of surface heat flux into the atmosphere, whose estimation requires knowledge of the spatial distribution of surface temperature.

Radiation

One or more components of the radiation balance at the earth's surface are measured at some stations; however, systematic measurements on a global scale are lacking. Satellite measurements can give us rough estimates of solar radiation, however. Although incoming solar radiation is a function of the optical properties of the atmosphere, the amount and type of cloudiness, and atmospheric profiles of temperature, humidity, and such major radiatively active gases as ozone and carbon dioxide, by far the most important variable is cloud cover. Current programs to map global cloud cover from satellites should fill an important gap in our current knowledge of the earth's radiation balance.

Sensible and Latent Heat Exchange

Sensible energy flux is not measured in any routine or systematic way, except in special experiments. The information is not much better for latent heat exchange. Except for some data collected in a few experimental settings and at a few lake evaporation stations, there are no reliable routine measurements available of either evaporative flux or the associated energy fluxes. In some countries, including the United States, efforts have been made to measure potential evapotranspiration by pans or other evaporimeters. Unfortunately, the relationship of these measurements to actual evapotranspiration is difficult to assess, because the devices are small, creating local anomalies of energy and humidity conditions, and because they do not incorporate the biophysics of plant behavior.

A field tool under development for remote three-dimensional measurements of water vapor fluxes between the surface and atmosphere is a system based on solar-blind Raman light detection and ranging (LIDAR). The technique involves repetitive firing of a short pulse of a high-intensity ultraviolet laser into the atmosphere and detection of the returning fluorescence as a function of time, with interpretation of backscatter intensity to determine water vapor concentrations. Recent developments in laser technology, hardware, and software suggest that the LIDAR system could provide a sensitive (5 percent), rapid measurement of water vapor concentration profiles with a spatial resolution of less than 5 m along the laser axis. Successful development of the LIDAR system would provide a quantum advance in estimating relative evapotranspiration in mixed vegetation, complex topography, and incomplete vegetative cover. It would also bridge the gap between point measurements of evapotranspiration and broad-scale estimates derived from satellite sensors.

Surface and Subsurface Data

Topography is known to adequate accuracy over most of Europe, Canada, and the United States, but topographic data are not easily available for most of the rest of the world.

Land cover is not routinely monitored at scales appropriate for hydrologic investigations.

Subsurface information on the hydrologic properties of geologic media is generally only available where aquifers have been intensively developed for water supplies.

Topography

Examination of hydrologic processes also requires other information about the surface and subsurface of the earth. Among these, perhaps the most important is topography. Elevation and related parameters (slope, aspect, drainage area) exert an important control on surface and subsurface hydrology and ecosystems. Topography influences intercepted radiation, precipitation and runoff movement of sediment, evaporation, soil moisture, and vegetation characteristics.

At a given latitude, the topographic parameters determine the exposure of a landscape to weather and sunlight and thus determine its microclimate. Local topography exerts a critical influence on microclimate through its control of the radiation balance, both solar and emitted atmospheric radiation. These in turn affect evapotranspiration, snow ablation, and the movement of soil moisture. Calculation of accurate energy fluxes in mountainous areas requires accurate information on topography, because the high local relief produces a complex array of microclimates and a complex distribution of vegetation.

Quantitative modeling of water, nutrient, and sediment transport and of nutrient cycling has been an important activity in hydrologic research. Human activities have affected the cycle of some biologically active substances, especially in aquatic ecosystems. Topography strongly influences surface and subsurface fluxes of water, but to analyze these accurately requires derivative quantities based on elevation data (local gradient, upslope area).

Digital topographic data enable automatic calculation of watershed structure and runoff, but error propagation associated with parameters, based on the first or second derivative of elevation, dictates that original elevation data must be of high quality. Many of the existing topographic data are inadequate, even in the developed parts of the world. Regional and global hydrologic studies require the manipulation of large data sets and therefore require topographic data to be accessed digitally. At present the availability of this type of data is limited. High-resolution digital topography (about 100-m horizontal resolution, showing about the same level of detail as a topographic map at a scale of 1:250,000 or better) exists only for the United States, Australia, and Western Europe. The best global coverage has only 18-km horizontal resolution.

In the investigation of global hydrologic problems, an important area of uncertainty is the mass balance of the great ice sheets in Antarctica and Greenland. Elevation changes experienced by ice sheets can be a response to climatic forcing or may be caused by internal instabilities in the ice flow, such as surges. Currently we do not know whether the polar ice sheets are stable, growing, or shrinking. Investigation of such problems depends on topographic data, not the data with fine spatial resolution needed for accurate calculation of slopes over individual drainage basins, but more widely spaced elevation measurements accurate to within a few centimeters.

Where available, topographic information is usually in the form of topographic maps, consisting of hand- or machine-drawn contours around unevenly distributed spot height measurements. For many parts of the world, coverage is limited, inaccurate, or nonexistent. Mountain belts, deserts, tropical rain forests, and polar areas, all critical environments for research on the earth's water balance, lack adequate topographic coverage. Even where topographic maps exist, they may have been generated in a way that limits their usefulness. For example, the distribution of horizontal and vertical control points across a map is usually uneven, and this results in variable accuracy.

Land Cover

Outside North America and Europe, most data on land cover are compiled from regional or national atlases, and the accuracy of such data is usually not known because human activity changes the surface cover. Extrapolation of measurements and development of models of hydrologic fluxes from field experiments conducted over typical surface covers to the broader region, requires that information on land cover be frequently updated.

The influences of vegetation on evapotranspiration into the atmosphere and on fluxes of carbon dioxide toward the surface are of special concern. Simulations of atmospheric processes and measurements of isotopic concentrations in rainfall show that precipitation over continental areas, and the temperature contrast between continents and oceans, together with associated monsoon circulations, are sensitive to the availability of moisture at the land surface. Under healthy, dense vegetation cover, soil moisture in the root zone is used freely by plants to maintain temperature control by evaporation, in association with leaf respiration and photosynthesis. If soil moisture is depleted, these fluxes are reduced. Today the best example of the need for this kind of information is in the Amazon basin, where some areas are undergoing a rapid transition from forests to clearings. Satellite measurements of forest clearing can be incorporated into large-scale hydrologic models.

Subsurface Information

Characterization of a ground water reservoir requires data on the extent, thickness, and structure of the geological units at and below the ground surface; on the hydraulic properties of each of the geological units; on the depth to the water table; on the chemical character of dissolved solutes within each unit; and on the areal distribution of ground water recharge. During the last decade, many countries have supported programs to document the hydrology of their principal aquifer systems. This task involves an integration of the geological, hydrologic, and geochemical data as a flow system analysis. A good example of this effort is the Regional Aquifer System Analysis (RASA) program of the U.S. Geological Survey. The motivation for much of this work is to provide information that can be used to better manage ground water resources. Such work also provides data useful in identifying aquifers from which society will benefit by using its resources to protect and perhaps restore the quality of water that has been contaminated.

SOME OPPORTUNITIES TO IMPROVE HYDROLOGIC DATA

Coordinated Experiments

Hydrologic models and field measurements in a coordinated program are needed to obtain a detailed understanding of the energy and water cycle at scales beyond that of the field plot.

Large-scale field experiments, combining satellite and extensive in situ measurements, are a means to verify large-scale hydrologic models and validate spatially extensive observations.

The essence of physical science is experimentation. To describe a physical phenomenon requires that it be considered at a given scale— the scale that is available (depending on the data) or a scale that is chosen (depending on the objectives of the study). It is now generally accepted that further progress in the development of the needed parameterizations of land surface fluxes of energy and water in climate dynamics and in hydrology must be based on a new, chosen scale, with comprehensive experiments conducted and field data collected at larger scales than those customary in the past. This recognition has led to cooperation at various national and international levels, resulting in several large-scale field experiments, which are either in progress or in the planning stages, to study the dynamic interactions of the land surface and atmosphere. These multidisciplinary experiments can often achieve more than the sum of their separate disciplinary goals if observations are coordinated to achieve a common objective. We refer to such efforts as coordinated experiments.

The design and execution of coordinated experiments constitute a landmark in hydrology. Several experiments are now being actively planned in the United States and internationally. Studies should be conducted in the future at different locations and under different climatic conditions, to validate existing concepts or to develop new ones that suit changed circumstances.

Scientific Development and Achievements

Awareness of the importance of experimentation has always existed, but it is only recently that it has become possible to consider experiments at scales that were unthinkable only a decade ago. New opportunities have arisen as a result of developments in low-cost electronic instrumentation, computer technology for handling large data sets, and remote sensing from satellites and aircraft for observations with appropriate spatial scales.

One stream of activity has developed from the realization that simulations of the earth's climate using general circulation models (GCMs) are sensitive to evaporation from land surfaces. Thus the idea ripened that pilot experiments are needed at a scale of about 100 km to provide good sets of data together with independent determinations of energy and water vapor fluxes near the surface, so that realistic boundary conditions can be put into GCMs. As a result, in the summer of 1986 coordinated observations, known as the Hydrologic-Atmospheric Pilot Experiments—Modelisation du Bilan Hydrique (HAPEX-MOBILHY), took place in southwestern France. The experiment was conducted between Toulouse and Bordeaux over an area that measures 100×100 km^2 and is about 40 percent forestland (the Landes Forest) and about 60 percent open agricultural terrain (Figure 4.6). Coordinated observations of variables and fluxes in physical state were made by satellite and from aircraft, sounding balloons, and ground stations. (See "Coordinated Field Experiments," Chapter 7, for more detail.)

A second line of experimental activity arose within the International Satellite Land Surface Climatology Program (ISLSCP). ISLSCP was organized in response to the need to monitor variables that govern climate and its fluctuations at different regional and global scales. Satellites are eminently suited for this purpose. A first objective of ISLSCP is to develop and use relationships between current satellite measurements and hydrologic and other climatic and biophysical variables at the earth's land surface. A second objective is to validate these measurements and relationships with ground data and also to validate surface parameterization methods for simulation models that describe

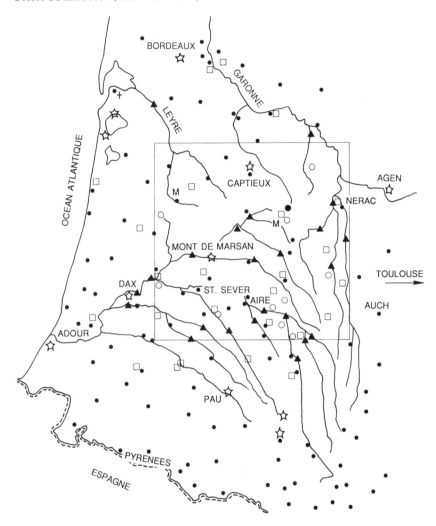

FIGURE 4.6 Location of HAPEX experiment in southwestern France. SOURCE: Reprinted, by permission, from Andre et al. (1986). Copyright © 1986 by the American Meteorological Society.

surface processes ranging from those at the scale of leaves of vegetation up to those at scales appropriate to satellite remote sensing, i.e., 100 to 1,000 m.

The First ISLSCP Field Experiment (FIFE) marked the initial phase of an experimental effort envisioned to accomplish these goals. FIFE took place during the summer and autumn of 1987 and 1988 near

THE GLOBAL ENERGY AND WATER CYCLE EXPERIMENT (GEWEX)

The Global Energy and Water Cycle Experiment (GEWEX), proposed to begin in the late 1990s, is designed to verify large-scale hydrologic models and to validate global-scale satellite observations. This initiative of the World Climate Research Program addresses four scientific objectives:

1. To determine water and energy fluxes by global measurements of observable atmospheric and surface properties;
2. To model the hydrologic cycle and its effects on the atmosphere and ocean;
3. To develop the ability to predict variations of global and regional hydrologic processes and water resources and their response to environmental change; and
4. To foster the development of observing techniques and data management and assimilation systems suitable for operational applications to long-range weather forecasting and to hydrologic and climatic predictions.

A central goal of the GEWEX program is to develop and improve modeling of hydrologic processes and to integrate surface and ground water processes on the catchment scale into fully interactive global land-atmosphere models. Inadequate representation of hydrology is a major weakness in present climate models. For example, a radical improvement is needed in the treatment of evapotranspiration, which dominates water and heat fluxes from the land surface.

The GEWEX program plans to support hydrologic modeling of continental-scale river catchments encompassing a diversity of terrain and climate conditions. The GEWEX field experiment program would systematically test these models on selected river basins for a minimum of five years, in order to provide the opportunity to compare detailed performances of alternative models under realistic conditions, to ascertain their sensitivity to different estimates of forcing fluxes, and to determine their agreement with observations. The experimental areas must be large enough that the hydrologic processes that contribute to global climate models and large-scale meteorological processes are apparent. The areas should encompass a wide range of soil moisture conditions, vegetation types, and surface topographies. Candidates would include the major river basins of the continents, for example, the basins of the Mississippi, Nile, and Amazon.

The GEWEX program is currently formulating an interim science plan supported by funding agencies in several nations. To carry out this work it will be necessary to sustain the momentum and the enthusiasm that characterized the early work in HAPEX-MOBILHY and in FIFE. Field studies are time-consuming and require special expertise that can only be developed through long and uninterrupted experience.

Manhattan, Kansas, over an area of about 15 × 15 km², which included the Konza Prairie Long-Term Ecological Reserve. The experimental area consists of rolling hills with about 50 m of relief between ridges and stream valleys, typically separated by distances of about 1 km. The FIFE study area is roughly representative of a much larger area, since it is surrounded by similar grasslands, which are mainly used for grazing. This area of tall grass prairie covers a strip 50 to 80 km wide that runs from Kansas to Nebraska to Oklahoma. The objectives of ISLSCP and FIFE imposed the need for simultaneous data acquisition and multiscale observation and modeling. As illustrated in Figure 4.7, this was done by acquisition of satellite data, together with si-

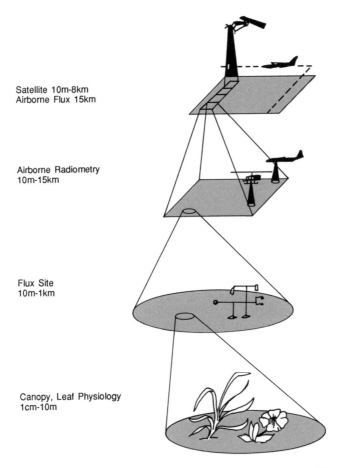

Satellite 10m-8km
Airborne Flux 15km

Airborne Radiometry
10m-15km

Flux Site
10m-1km

Canopy, Leaf Physiology
1cm-10m

FIGURE 4.7 Range of scales addressed in First ISLSCP Field Experiment (FIFE). SOURCE: Courtesy of P. J. Sellers, University of Maryland.

THE NATIONAL STORM PROGRAM

The current state of understanding of meteorological processes can be summed up by the paradoxical state of weather forecasting. We have improved the prediction of large-scale weather events substantially in recent years, such that a three- to five-day outlook for a sunny weekend, or for a wet period, is quite likely to be correct. And yet, short-term prediction of precipitation and severe weather events, although more accurate than in the past, has not kept pace. Snow flurries in today's forecast can still give way to paralyzing blizzards, and scattered showers anticipated for this evening occasionally become deluges that can and do cause devastating flash floods. Less dramatically, forecast rains may appear only in adjacent areas while sunshine returns unexpectedly, making millions of people wish that they had carried out outdoor activities, after all.

We now can explain this state of affairs, and in so doing, understand why the National Stormscale Operational and Research Meteorology (STORM) Program will be an important part of the 1990s. We have improved prediction of large-scale events because our understanding, observational capabilities, and computational capabilities are equal to the task. The atmospheric processes that control the weather on time scales of days have spatial scales of thousands of kilometers, and for the most part their behavior is well handled by current numerical models and computers. But we now know that virtually all precipitation and severe weather events involve mesoscale processes (literally, intermediate in scale between large-scale cyclones, anticyclones, and planetary waves, and short-lived phenomena such as turbulent eddies or individual clouds), which are often tens to hundreds of kilometers in extent and last from hours to a day or two. Our current observational network fails to describe such events, just as a whale net fails to catch most fish—the mesh is too large. But we are moving rapidly toward understanding such systems, and new observational capabilities and more powerful computers will greatly improve prediction, in turn greatly mitigating the human and economic impact of these events. The conviction that mesoscale weather events are both comprehensible and predictable is the central premise of the National STORM Program, a multiyear program of both operations and research.

The goals of STORM are to (1) advance fundamental understanding of precipitation and other mesoscale processes and their role in the hydrologic cycle, and (2) improve the 0- to 48-hour prediction of precipitation and severe weather.

To achieve these goals, STORM builds on two extraordinary national investments in weather and climate instrumentation and associated information technology, each scheduled for deployment and operation during the 1990s: the modernized national weather observing system, and the Earth Observing System.

The new generation of observing technology, in particular, has been designed specifically for observing mesoscale weather events. Doppler radars have been used for years in research but will soon be operational throughout the country. Information on the time-space density of wind, temperature, and precipitation will increase by over an order of magnitude. However, an investment in technology alone is not enough. A second commitment is required: to sustain advances in scientific understanding. STORM represents the national strategy for achieving the needed level of understanding of precipitation and other mesoscale processes and combining that understanding with the immense dollar investments in hardware for the benefit of the nation. While initially stimulated by the benefits to be gained from improved weather prediction, STORM will profoundly affect our understanding of that most important element of global change: the hydrologic cycle. Mesoscale weather events, including severe events such as hurricanes, mesoscale convective systems, and winter storms, dominate the climatological averages that describe worldwide precipitation. The practical effect of global warming is not made manifest as much by increased global average temperature as it is by the uncertain extent of changes in regional and seasonal precipitation around the globe and the implications for the hydrologic cycle.

If they can be separated at all, hydrology and the National STORM Program are interdependent. For STORM to be successful, hydrologists must participate in fundamental research on the structure of rain systems and on the factors determining evapotranspiration from land surfaces, among many other basic research problems. Combined meteorological and hydrological models are required to bring the benefits of improved understanding and observations of precipitation to realization in better flood forecasting and warnings. Conversely, the operational and research data base provided by STORM data sets will be most valuable in advancing fundamental knowledge of near-surface hydrology, and its characterization by remote sensing, during the next decade.

multaneous observations from aircraft and at numerous ground stations, and by atmospheric soundings of different types.

As international efforts both HAPEX-MOBILHY and FIFE posed severe and formidable problems of logistics and coordination. One of their major accomplishments has been attaining and sharpening experimental technology and expertise for use in planning later experiments.

Frontiers and Challenges

It is important for future experiments that address the interaction between surface hydrology and the atmosphere to increase the area examined to the regional scale. Possibilities under discussion for the 1990s include a combined experiment with the National Stormscale Operational and Research Meteorology (STORM) Program and an ISLSCP experiment that includes interactions between hydrology and terrestrial ecology, probably in the boreal forests of the United States and Canada. The STORM program will be primarily a meteorological effort dealing with mesoscale-storm-generating mechanisms, but it should have profound ramifications for hydrology as well. These future ISLSCP experiments are envisioned to involve scales of 200 to 300 km, the grid size of GCMs, and would be best conducted along a gradient of different forest species. Because it will be impossible to measure all relevant variables over such large areas, several experiments of the spatial scale of FIFE will be nested within the larger area. It will be necessary to design experiments to tie together some of the findings of experiments at the level of HAPEX, ISLSCP, and STORM to the continental or even global scale. The Global Energy and Water Cycle Experiment (GEWEX), proposed to begin later, around 1997, should achieve this objective.

Remote Sensing

Over the next decade, advances in remote sensing—the gathering of data by instruments on satellites, aircraft, or the surface to infer properties of the subsurface, surface, and the atmosphere—offer the possibility of obtaining frequent hydrologic measurements over wide spatial scales.

To achieve the potential benefits of remote sensing, the data must be converted from the raw electromagnetic measurements made by satellites, by aircraft, or at the surface to hydrologic information that is made available to a wide spectrum of hydrologic scientists.

The hydrologic sciences, much like the rest of modern earth science, are starting to examine interactions among the different terrestrial components at all temporal and spatial scales. Such an inclusive perspective requires an integrated data collection program, and remote sensing is an essential component. Instruments currently available on satellites, aircraft, or on the surface, along with those planned for the future, will make available measurements obtained throughout the electromagnetic spectrum over a range of spatial and temporal scales.

The temporal variations of many hydrologic processes require global coverage every few days, and so satellite instruments with broad swath widths and modest spatial and spectral resolutions are necessary. In addition, some hydrologic problems require analysis and interpretation of specific areas and detailed sampling within scenes produced by instruments with lower spatial or spectral resolution. For these applications, data from satellite or aircraft instruments with appropriate higher spatial and, usually, higher spectral resolution are needed. Some processes that are important at the global scale are manifested in surface features with dimensions of tens of meters. Examples include anthropogenic damage to vegetation, which first appears in patches; forest clearing, which dramatically affects evaporation and carbon cycling, and which in the tropics occurs in small, noncontiguous areas; land use change and desertification, where boundaries may move only short distances; alpine snow and ice, where spatial coverage may be small but where large volumes of water are stored; changes in permafrost and buried ice lenses caused by atmospheric warming; and changes in the extent of freshwater and saltwater marshes caused by changes in the water table height or sea level.

Remote Sensing of Hydrologic Parameters

Some important hydrologic variables can be measured by remote sensing. In the visible and near-infrared wavelengths, the source of energy is the sun, and we can measure the solar radiation that is reflected by the surface or scattered by the atmosphere. In the infrared wavelengths, we measure radiation that is emitted by the earth and its atmosphere. In the microwave part of the spectrum, we measure either emitted radiation (passive microwave) or the backscattered response to a signal sent from a satellite or aircraft (active microwave). The visible and near-infrared wavelengths can be used to measure the presence or absence of vegetation; the structure of vegetation, including biomass and leaf density; the stress in vegetation, including moisture content of leaves; soil type; and snow cover and its rate of depletion.

For hydrology, the microwave region offers particular advantages, because the signal is integrated over some depth below the land surface, whereas reflectance of solar radiation and emission of thermal infrared radiation are determined by the characteristics of a much thinner surface layer.

LOUIS J. BATTAN
(1923-1986)

Some of us struggle through life to find the right career path, while others take a seemingly random step that sets a route early. Louis J. Battan, the New York City-born son of Austrian immigrants, achieved his renown as a meteorologist because of an Army recruiting poster. Although he began college as a mechanical engineering major and passed the New York City fire department entrance exam in case he ever needed a backup career, it was a World War II recruiting poster that led him into the Army Air Corps aviation cadet program in meteorology.

Battan was sent first to New York University to train in meteorology. Finishing in the top 10 of his class, he was selected by the Army as one of the privileged 100 weather officers to be trained also in radar at Harvard University and the Massachusetts Institute of Technology. His war-time duties—including ground-based storm detection and airborne weather reconnaissance—set a strong foundation for his later life's work, and he often credited his dual training for giving him an ability to see problems with an innovative eye.

After the war, Battan earned a meteorology degree from New York University. He then joined the U.S. Weather Bureau and was assigned to the Thunderstorm Project at the University of Chicago, where he later earned both his master's and doctoral degrees. The Thunderstorm Project was the prototype for many large-scale field experiments following the war. It involved the coordinated use of radar, an extensive ground network of meteorological stations, and storm-penetrating P-61 Black Widow aircraft, and the project brought a giant leap in our understanding of storm behavior. Battan achieved the first definitive identification of the coalescence mode of precipitation growth in warm convective clouds. Also during this time, he worked with others on the artificial nucleation of cumulus clouds. A main theme in his research was the radar backscatter from hail. He also was a pioneer in the development of Doppler radar. Throughout his career, he stayed active among his various interests—weather modification, Doppler radar, and the scattering properties of hydrometeors.

A student of language—he spoke Italian, Spanish, French, German, and Russian—Battan was by choice a life-long university professor and writer. He authored some 16 books during his career, and literally hundreds of papers and articles. Although offered many high positions in government and academia, he preferred his scholarly lifestyle. He served on many national and international geophysical committees—including the National Academy of Sciences' Committee on Atmospheric Sciences, which he chaired—and was a delegate to the World Meteorological Organization. He was influential in founding the National Center for Atmospheric Research. Battan, always at heart a scholar, fought often for the right of scientists to guide their own fates without undue direction from above, believing that the best science is done in an unfettered, supportive atmosphere.

EOS—NASA'S EARTH OBSERVING SYSTEM

The first polar platform of NASA's Earth Observing System (EOS) is scheduled for launch in 1998. This mission, planned for a duration of 20 years, will provide geophysical products from innovative spacecraft sensors. To improve our knowledge of earth system science, including the hydrologic cycle and its interaction with the physical climate and biogeochemical cycles, EOS has three distinct goals:

1. Establishment of long-term, reliable measurements from remote sensing of important hydrologic variables, so that hydrologic change over the two decades after launch can be documented;

2. Use of remote sensing data, from EOS platforms and from aircraft and other satellites, to identify and investigate the most important hydrologic processes; and

3. Improvement of our predictive models, so that plausible hydrologic change over the next century can be better understood.

The complement of instruments scheduled for EOS includes many that are important to the hydrologic sciences. Particular examples include:

• an infrared sounder for atmospheric temperature, water vapor, and trace species;

• a microwave radiometer for atmospheric water vapor, precipitation, and snow and ice extent in all weather conditions;

• two imaging spectrometers—one at high spatial resolution for detailed investigations, and the other at coarser resolution for global mapping—for measurements of nutrients in oceans and coastal and inland waters, optical properties of snow and ice, and tropospheric water vapor; and

• a synthetic aperture radar for all-weather studies of small- and large-scale structural characteristics of the surface, particularly useful for measurements of glaciers, soils, and vegetation.

The EOS Data and Information System (EOS DIS) will provide timely data to investigators at the marginal cost of reproduction. Some of these experts will analyze the data to provide hydrologic products, and these products will also be available in EOS DIS for use by other investigators. In this way, EOS will open the capabilities of remote sensing data to a broader range of scientists, because they will no longer need detailed knowledge of instrument characteristics and electromagnetic interactions at the surface.

Because clouds do not obscure the microwave signal over much of the wavelength range, remote sensing from aircraft or satellites is possible even when cloud cover is present. Moreover, in the microwave frequencies the water molecule is resonant, and the electromagnetic properties of wet substances are much different from those of dry ones. Emitted (passive) or reflected (active) microwave radiation is especially suitable for measurement of rain rate, from ground-based radar; estimation of soil moisture; measurement of hydrogeologic properties, from ground-penetrating radar; and mapping of snow cover.

Many categories of hydrologic variables can be measured by remote sensing, because hydrologic processes modify the electromagnetic signal in some portion of the spectrum. However, different hydrologic conditions may cause similar signals, and so continued work is required to achieve unambiguous measurements. We need progress in two categories of problems:

• We need to understand better the relationship between properties of the surface and its electromagnetic signature; and

• The model between the electromagnetic signature and the physical properties may be complicated, and the inversion of such a model, so that surface properties can be estimated, is often difficult.

Future Advances

New satellite systems will open uncommon opportunities for hydrologic research in the coming decade. In the near term, before 1992, these systems include (1) the DMSP Special Sensor Microwave Imager with 12.5- to 25-km resolution, the first launched in June 1987 and later launches scheduled to continue into the 1990s, and (2) the ESA ERS-1 with synthetic aperture radar (SAR) and altimeter, to be launched in 1990. Additionally, NOAA polar-orbiting satellites of the 1990s will have additional channels for the advanced very-high-resolution radiometer (AVHRR) in the near-infrared wavelengths and an advanced microwave sounder. In the longer term, the Earth Observing System (EOS) scheduled for launch in 1998, with its full complement of instruments and the ambitious plans for research across the earth sciences, will allow rapid access to data over global and large regional scales for the purpose of tracking the changing hydrologic cycle. In addition, the Tropical Rainfall Measuring Mission (TRMM) will improve the knowledge of rainfall over important tropical areas of the earth where data are scarce.

One important focus of these future missions is the creation of hydrologic data products. A current impediment to the use of remote sensing in hydrology is the investment of time necessary to understand the characteristics of sensors and the relationship between electromagnetic and hydrologic properties of the surface, along with the cost of data and difficulty of access. Scientists with little previous experience in using remote sensing data are often discouraged by the amount of technical expertise and the sophistication of the computer equipment required to process the data. In the future, hydrologic scientists who understand these kinds of problems must create data products of established quality and reliability, so that other scientists can use these in their models. The hydrologic community should start now to be prepared to use these data products, which will cover large areas of the earth at a frequency of one to a few days, and smaller areas in more detail at intermittent sampling frequencies.

Opportunities for Effective Use of Current and Planned Sensors

The prospects are promising for remote sensing in hydrology. The sensors designed for future spacecraft missions have excellent characteristics for measurement of hydrologic properties. However, data availability and distribution are causes for concern, as are defining and processing suitable data products. Use of remote sensing in hydrology, as in other disciplines, has been hampered because the data are difficult to acquire and analyze, and some data are too expensive. The best scientific results with remote sensing data are usually achieved by analysis of multitemporal as well as multispectral data; therefore data need to be priced so that investigators can analyze many images during a single season and thereby observe both spatial and temporal changes. Two improvements in the current state of hydrologic education and research are required to achieve the potential benefits of remote sensing data.

1. Universities must recognize that students in the diverse array of disciplines that analyze the earth's hydrologic processes will need to be trained in remote sensing, as well as in the conventional supporting subjects of physics, chemistry, mathematics, and computer science. Departments that offer courses and degrees in hydrologic science will need to incorporate remote sensing training into their curricula.

2. The hydrologic community will need to agree on the design and distribution of hydrologic information products.

The future of hydrologic remote sensing will be enhanced by the wide range of hydrologic information that these data can produce. The instruments that are planned will contribute to science, because they can be used to measure many useful surface properties for hydrologic research, but the contribution will be greater if the data become more widely available in useful products, and if the population of hydrologists familiar with remote sensing increases. Specific recommendations are the following:

1. The continued availability of data from current sensors is necessary for hydrology. Landsat, for example, provides a time series dating back to 1972, and the continued viability of this system in the future must be assured.

2. The continuity between current and future sensors must be ensured, so that the time series created are long enough to interpret hydrologic change.

3. Future sensors appropriate for hydrology need to be designed, constructed, and launched. Each of the facility instruments for the polar platform of NASA's EOS contributes to hydrologic research.

4. Remote sensing data must be distributed soon after acquisition, and data must be available at regular intervals so that both spatial and temporal distributions of hydrologic phenomena can be investigated.

5. Hydrologic data products must be defined and processed from raw remote sensing data, so that the information will be available to a wider variety of scientists instead of being restricted to those with expertise in remote sensing. The definition of the most appropriate hydrologic products and the procedure for selecting the best algorithms will need to be addressed in a systematic way, perhaps by workshops that would bring together an appropriate mix of hydrologists and experts in remote sensing.

6. Hydrologic interpretations from remote sensing must be checked against surface observations.

Remote Sensing Below the Surface

Research frontiers in remote sensing of subsurface conditions promise important breakthroughs in our capability to map hydrogeologic properties without the need for a dense network of boreholes. Of particular note are techniques based on ground-penetrating radar and tomographic reconstruction. Ground-penetrating radar can provide high-resolution maps of the subsurface stratigraphic profile to depths of tens of meters. Work is under way to determine how a radar profile can be interpreted to characterize the subsurface structure of hydraulic conductivity, the key physical property determining patterns and rates of fluid flux. Using ground-penetrating radar, it may also be possible to delineate plumes of contaminated ground water.

Borehole tomography is similar in concept to the familiar CAT-scan equipment used in medical applications to obtain "pictures" of body organs without surgery. For applications in hydrogeology, multiple cross-hole signals are generated by the movement of a source along one borehole, while receivers record signal arrival at multiple depths in adjacent boreholes. Various source signals are being developed, including seismic, electromagnetic, and hydraulic. The cross-hole signals are processed to reconstruct a picture of hydrostratigraphic properties throughout the region between the source and receiver boreholes. Basic research is needed to better understand the relationship between the recorded signal and the hydrogeologic properties of the subsurface. If the potential of these two methods is realized, they will herald a new era in detailed mapping of subsurface properties relevant to hydrologic processes.

Isotope Geochemistry

Environmental isotopes can be used as tracers to study residence times, mixing ratios, and flow velocities in the hydrologic cycle.

Environmental isotopes are a key tool in studying the subsurface component of the hydrologic cycle. Their primary uses include:

• identification and differentiation of water masses that have unique mixtures of different isotopes of hydrogen or oxygen in the water molecules;
• determination of the extent of mixing of two or more waters;
• estimation of residence time in hydrologic systems; and
• estimation of flow direction, travel time, and flow velocity.

The environmental isotopes in most common use today include the stable isotopes deuterium (^{2}H), oxygen-18 (^{18}O), and carbon-13 (^{13}C), and the radioisotopes tritium (^{3}H), carbon-14 (^{14}C), and radon (^{222}Rn).
Because the stable isotopes oxygen-18 and deuterium are part of the water molecule and are not significantly affected by chemical interaction with the rock or soil, they provide nearly ideal conservative tracers of water masses. Owing to differences of mass between hydrogen and deuterium, and between oxygen-16 and oxygen-18, one isotope is enriched (fractionated) compared to the other as water changes phase from liquid to vapor to liquid to ice and back. Isotopic frac-

tionation is a function of temperature, and well-documented effects cause enrichment or depletion of one isotope compared to the other, allowing quantitative evaluation of processes occurring within the hydrologic cycle. For example, winter precipitation is depleted in ^{18}O and ^{2}H compared to summer precipitation. Distinct differences occur in the stable isotopic composition of precipitation with changes in latitude and altitude, and so changes in the trajectory of storm tracks cause changes in the isotopic composition of precipitation. Meteoric water that has experienced evaporative processes is enriched in ^{2}H compared to ^{18}O in comparison to normal seasonal variations observed in precipitation. On a smaller scale, rain that falls in the beginning of a storm is often enriched in ^{18}O and ^{2}H compared to that falling at the end of a storm. Isotope hydrologists can take advantage of these differences in applications such as the following:

• determination of the extent of leakage between aquifers;
• identification of recharge areas;
• determination of recharge rates or water age by counting annual cycles in ^{2}H and ^{18}O moisture in the unsaturated zone;
• application of isotope mass balances to determine the extent of interaction between rivers and aquifers; and
• investigation of paleoclimatic conditions.

Incorporating isotopic data with new advances in geochemical and hydrologic modeling promises considerable potential for gaining new insights into subsurface hydrologic processes and their link to processes occurring on the land surface.

During the period from 1957 to 1963 massive amounts of tritium were introduced into the upper atmosphere from the testing of thermonuclear weapons. In 1963 and 1964 the tritium content of precipitation reached as much as a thousand times natural background levels and established a unique tracer in ground water systems. Recharge rates have been determined by locating the depth to the 1963 tritium spike. The simple presence of detectable tritium is an indication that a sample contains post-1957 recharge water. Tritium decays with a half-life of 12.3 years, and dating of shallow ground water is possible if the tritium input to the hydrologic system can be defined (Figure 4.8).

Many other environmental isotopes and transient atmospheric tracers are being studied to investigate their possible applications as dating tools and tags of water masses in hydrologic systems. For dating techniques, research is directed toward the use of chlorine-36, silicon-32, argon-79, krypton-85, krypton-81, helium-3, helium-4, and fluorocarbon compounds. Heavy isotope ratios such as strontium-87/strontium-86 and uranium-234/uranium-238 can be applied as tracers of water

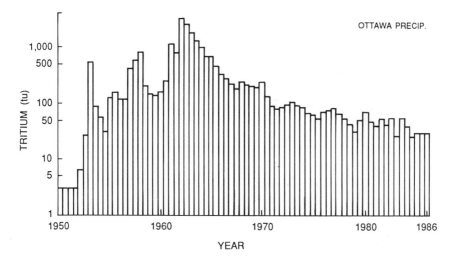

FIGURE 4.8   Tritium concentrations in precipitation measured at Ottawa, Ontario. SOURCE: Reprinted, by permission, from Robertson and Cherry (1989). Copyright © 1989 by the American Geophysical Union.

masses. Stable isotope ratios of sulfur and nitrogen often show source material and currently are applied to studies of anthropogenic inputs of sulfur and nitrogen into shallow ground water and surface water, such as from acid-rain deposition, agricultural runoff, fossil fuel combustion, and sewage sources. The observed distribution of the decay series of naturally occurring uranium-238 and thorium-232 can be used to derive sorption rates and retardation factors for use in site-specific models of the transport of radioactive or stable elemental waste in saturated geological media. These and other chemical and isotopic tracers hold great potential for opening fruitful avenues of research.

Paleohydrology and Long-Term Records

Paleohydrologic information can extend the time series of hydrologic data beyond the period of record and thereby give us a better picture of hydrologic trends and the statistical distribution of phenomena.

To understand hydrologic processes one must know how the processes vary through time, from the minutes of a cloudburst and hours of a flash flood through to the variations in precipitation over decades, centuries, millennia, and even longer. Neither means nor variations are necessarily constant over time. Within decades and centuries there can be extreme events, abnormal periods, and even climatic shifts. Detection and quantification of extreme events and climatic shifts are crucial for hydrologic evaluation. Traditional monitoring is useful in many applications, but new methodologies or increased efforts made with established procedures are necessary to produce information about long-term effects of climatic variation on hydrology. A current opportunity to improve hydrologic data is the development of new long-term records of climatic-hydrologic variation. These are useful for process modeling and making predictive or probabilistic estimates that may be beyond the range of recorded events. Important knowledge can be gained through paleohydrology—intensive study of historical documentation and generation of new data using proxy climatic-hydrologic information.

Paleohydrology can produce long-term information on means, extremes, trends, and variations of various hydrologic phenomena and probabilities of extreme events. For example, tree-ring analysis has led to reconstruction of records of precipitation, streamflow, drought, temperature, and lake levels. However, there are currently few of these reconstructions. Thus the feasibility has been established, but the extensive application awaits. There is a potential for reconstructing flood records, inundation times, sedimentation rates, ground water recharge, and other important phenomena relating to hydrologic hazards and water resources management. In addition, this type of analysis can identify and date single major events that may have occurred outside the realm of human records. Proxy records may be episodic, such as the identification of a major flood event in fluvial sedimentary deposits, where there may be organic debris for carbon-14 dating. Proxy records can be sequential, such as tree-ring or ice-core data, from which a continuous time series of variation can be recovered. Continuous, accurately dated time series can be used just like measured records to calculate means, variations, extremes, and shifts over time. Another application of these time series is the study of periodicities. Some scientists believe there is a response in climate variation to the 18.6-year lunar-nodal periodicity. One needs almost 60 years of good data to have at least three complete cycles for examining and testing the 18.6-year hypothesis. In most areas of the world such information is unavailable.

The proxies that record and represent hydrologic variations are biological and geological entities that persist over time and show responses to environmental change. Tree-ring analysis and palynology are two primary biological methods of deriving paleoclimatic-paleohydrologic records. Tree-ring analysis provides absolutely dated, year-by-year records for the past several centuries and in a few cases for several thousand years. Palynology, the analysis of pollen and spores, provides a longer record but does not give either the resolution or the precise dating of tree-ring analysis. Geological recorders are varves, fluvial deposits and other sedimentary records, geomorphic changes, lake fluctuations, paleosols, ice cores, and geochemical sequences. The dating of most geological recorders is approximate, being based on radiometric, stratigraphic, or other techniques, but the information provided by the geological recorders may cover time spans of thousands to millions of years. Figure 4.9 presents the various paleohydrologic indicators and the time spans and resolutions derivable from them.

There are limitations to paleohydrologic indicators. Ice-core sampling sites may be the most limited in distribution, and pollen may be the most ubiquitous, although pollen must be deposited coherently by some continuous sedimentary process for best results. Tree-ring sampling opportunities are widely distributed over land masses except for deserts and polar regions, but are particularly plentiful in tropical areas. The geomorphic-geologic sources of information, such as fluvial flood deposits, are sporadic in distribution, as are paleosols, varved sediments, and paleolimnological opportunities. In areas where proxy paleohydrologic indicators are absent, geochemical indicators, including isotopic analyses, can potentially determine sources, ages, and histories of ground water.

Long-term records can be developed also by the study of ancient documents that have only recently been appreciated for their scientific value. Improvements in travel and communication are increasing the awareness of documentation relevant to climatic or hydrologic variation. From records of shipping transit times and sea-ice conditions to descriptions of famine and even reports of tax yields, one can glean information about effects of climate. These types of analyses require that linguists and historians be involved in the study of hydrology.

Caution is called for when utilizing such reconstructions, because the analogs on which proxies are based are often incomplete in critical secondary responses. Consistency checks should be made of the reconstructed data such as by using them in models that test the water balance, the timing of extremes, and so forth.

Humans now may be on the threshold of causing climatic change, with attendant disturbances to the hydrologic cycle. Some of these modeled and projected shifts have analogs in the past. Precise knowledge

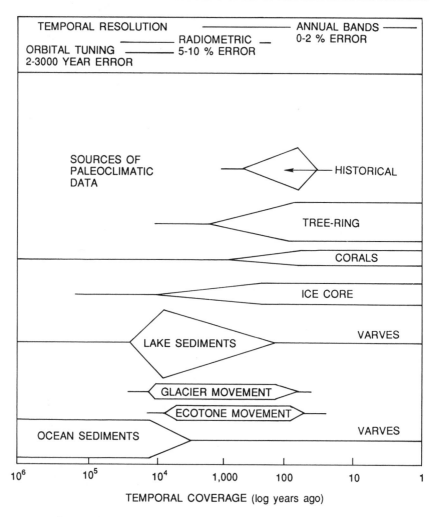

FIGURE 4.9 Paleohydrologic indicators and their time spans and resolutions. SOURCE: Modified from Bernabo (1978) courtesy of the U.S. Department of Commerce.

of the analogous conditions aids in planning for future eventualities. The measured records do not extend back to these conditions, but through examination of the different types of climatically sensitive organisms or cyclical phenomena that have left measurable responses to climate, we can identify major past events and develop time series that are proxies for measured climate variations. Global patterns of warmer or colder and wetter or drier regions can be ascertained through

---

PALEOHYDROLOGIC INFORMATION FROM TREE RINGS

Construction of a dam and allocation of water from the resulting new reservoir depend on an accurate prediction of the expected water yield. However, it is difficult to estimate the statistical properties of streamflow without a long time series. The available period of record may or may not include extreme events, abnormal periods, or climatic shifts, but the hydrologic data can be improved by intensive study of proxy records, which can yield long-term information. Such a reconstruction is called paleohydrology, and useful proxy information can be found in tree rings. Such data could have been used to more accurately estimate the statistical distribution of future flows on the Colorado River, before the water from Hoover Dam was allocated.

In 1922 the Colorado River Compact allocated at least 19.4×10^9 m^3 (15.75 million acre-feet) average flow per year at the Lee Ferry measuring point between the upper and lower basins. Based on data available at the time, the analysis estimated a mean annual flow of 18 million acre-feet available for allocation. As the bottom plot in Figure 4.10 shows, however, this allocation was based on the wettest period in the 400-year history reconstructed from tree-ring analysis. Information provided by paleohydrology, in this case tree-ring analysis, can allow for better judgments in future allocation decisions and show the frequency of low-flow periods so that drought probabilities can be estimated. Long-term data and paleohydrologic information are crucial for determination of hydrologic variations or shifts that are outside the range of the instrumental records.

---

study of biological and sedimentological indicators. Global circulation models, which can be run forward in time to produce estimates of future variations, can also be set back in time to enable comparisons with proxy data. Such exercises should validate the model mechanisms and concepts and also estimate potential changes in the global hydrologic cycle. The extent of the normally recorded information is too limited in time and space for such validation.

As a contributing discipline, paleohydrology should be considered still in its early stage. Reconstructions have been published for streamflow of specific rivers in the southwestern and eastern United States and in Argentina based on tree-ring analysis. This technique has also produced multicentury drought histories for certain regions along with calculations of return times for severe drought, at scales of major drainage basins. However, there are many key hydrologic uncertainties and data gaps throughout the world where tree-ring analysis and

264

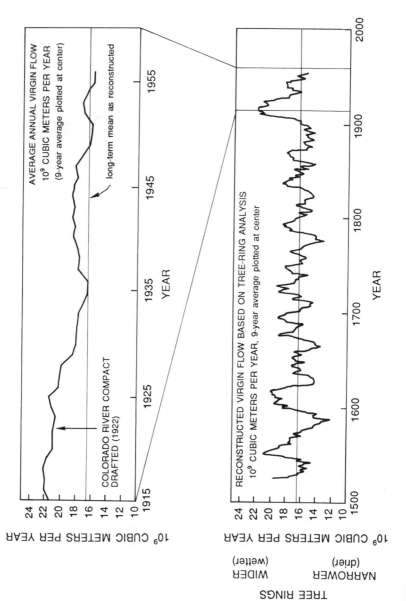

FIGURE 4.10 Tree-ring reconstruction of flow in the Colorado River. SOURCE: Reprinted, by permission of G. J. Jacoby, from Stockton and Jacoby (1976).

other proxy studies could be applied. Sometimes they provide information where there are no instrumental records even today, for example, in coral reefs and ice cores. To execute the application requires interdisciplinary science. There must be expertise in the particular proxy area—e.g., palynology, dendrochronology, sedimentology—and expertise in hydrology itself to develop sensible and significant results. Although it has been used successfully in many areas of the world, there are now more opportunities than accomplishments in paleohydrology.

Data Accessibility and Management

Advances in the hydrologic sciences depend on how well investigators can integrate reliable, large-scale, long-term data sets.

Storms, floods, and droughts are natural events that can be measured just once, whereas laboratory experiments can be repeated. Instruments used in hydrology must be reliable and operated such that data captured are of known standards and precision.

On rivers, measured stage data must be transformed to discharge. The stage-discharge relationships, commonly called rating curves, typically must be extrapolated to extreme stage values and may require adjustment as new gagings at the extremes become available. This adjustment may apply retrospectively for rating curves that have been used for many years, and so a data archive should store original stage measurements and rating curves separately, to allow this retrospective examination and adjustment.

The data sets required to answer many of the open research questions in hydrology will be complex. Inevitably, many scientists from a variety of disciplines and backgrounds will be involved in data collection and analysis, over a significant period of time. How can diverse investigators and investigations produce compatible data sets, assure their quality, and confidently assemble them for larger, indeed public, use and access? Creating effective data systems for assembling and distributing scientific data sets is not trivial and depends heavily on the personal efforts of active scientists. If the data systems are constructed within the scientific community by scientists themselves, rather than by independent data "experts," there will be many scientific opportunities as well as technical and political challenges.

Data Management in the First ISLSCP Field Experiment

The First ISLSCP Field Experiment has wrestled with these issues. FIFE was not designed exclusively as a hydrologic experiment, but it included collecting a comprehensive set of hydrologic and related data sets for a 15 × 15 km^2 experimental site in central Kansas. The studies of some 30 cooperating principal investigators were actively supported by a data system built by participating staff scientists at NASA's Goddard Space Flight Center.

The immediate goals of the data system were to capture and preserve the data and distribute them as rapidly as possible. After the conclusion of the field sampling, the system was converted into an open, widely available archive. The technical elements of data distribution were easily supported: magnetic media via mail for large volumes and an on-line access via electronic networks for browsing and routine data extraction. Equally important, however, was a user support staff. Technical and scientifically competent people were required to simplify access, prepare adequate documentation, and teach novice users about both the system and the data.

Assessing data quality can be difficult enough for a single investigator working with his own data. The problem is compounded when the data are required quickly by cooperating investigators and distributed through a data system to people who may be unfamiliar with the technical details (or difficulty) of the measurement. The FIFE solution was to use the data system as the focus for a cooperative assessment of data quality.

Data Storage and Access

Optical disks and compact disks have become an attractive alternative to traditional magnetic tape or disk storage media, because they offer the capacity and security necessary for hydrologic archives, and because multiple copies of large archives can be made cheaply. For example, the entire daily stream gaging record of all gaging stations for one year is stored on optical disks for such countries as the United States and New Zealand. The need to publish expensive yearbooks of data disappears.

Issues to Resolve

The evolving requirements characteristic of active research demand a data system with real-time adaptability. This can be achieved by a scientifically involved data system team that puts a priority on service.

NEXRAD: A NEW AGE FOR PRECIPITATION DATA

Most meteorologists and hydrologists who work with rainfall data have had little choice but to accept the deficiencies of data sets acquired for operational purposes. Recording rain gages are often widely spaced and not placed optimally within watersheds. Use of operational radar network data for rainfall estimation has been severely limited: the technology dates from the 1950s, the radars are difficult to calibrate, the beam width is too large for quantitative use except at short range, and the data are archived at too coarse a resolution for most research uses. Dramatic changes are about to take place with the coming of the next-generation weather radar (NEXRAD) system.

For the first time, beginning in 1990, quantitative precipitation estimates in digital form will be available both in real time and retrospectively for research. By the mid-1990s well over 100 NEXRAD stations will be in place, covering most of the United States. While these data will be subject to uncertainties, such as that resulting from the relationship between radar reflectivity and precipitation, imperfect beam filling, and shielding by topography, the NEXRAD data will greatly surpass current operational radar data in quality, areal coverage, and both spatial and temporal resolution.

Is the hydrologic science community prepared for the opportunities created by this approaching deluge (of data)? This is an urgent question, because without special arrangements only a small fraction of NEXRAD data may be archived for research use. There will naturally be operational use of the data in hydrometeorology, for example, for flash flood warnings.

Questions hydrologists should ask, for example, include: What will be lost if only hourly accumulation and 10-km-resolution data are archived, when data are potentially available every five minutes and with a spatial resolution of 1 km? Are there particular watersheds or other areas with hydrologic networks where long, high-resolution time series are required? Perhaps most importantly, are there important scientific questions that have not been addressed simply for lack of such quantitative precipitation estimates? The national investment in NEXRAD will be about $1 billion, and while it is designed primarily for improved storm warnings, it opens up the possibility of an unparalleled data base for hydrologists.

This is no small effort: it requires direction on a day-to-day basis by active scientists. In particular, for large projects, the role of a project information scientist must be recognized as critical and must be rewarded appropriately.

In addition to personnel issues, there are several political aspects. Three legal categories of data can be defined: data that are acquired from public sources with no distribution restrictions, data that are collected by publicly funded principal investigators, and data that are acquired with public funds from private sources, corporations, or individuals with specific legal rights to restrict their distribution.

The FIFE approach was to recognize a data collection and analysis phase in which data sets were exchanged and revised and their quality controlled, but during which general access was restricted. The experience suggests that the more direct control scientists have over the data system, the better it will serve science. In particular, direct control makes possible rapid adaptive responses to the unexpected opportunities that can develop in any experiment. A control data base can be a tool and focal point for cooperatively assembling and checking the data sets. Advances in computer technology make it possible to link such a data base electronically to field sites and investigator laboratories, where a common set of hardware and software tools can be inexpensively supported. These would allow each scientist to create his or her portion of the data base, in near real time.

Challenges in Measuring Water Quality

Public concern with pollution of water resources, as well as its effects on human health and the environment, is widespread and occasionally intense.

Investigations of water quality must be designed according to sound scientific principles.

In response to public concern, many studies are being conducted to monitor and assess the amount and distribution of pollutants entering the hydrologic cycle. If these studies are to be useful to understand the causes of observed conditions, and thus provide a foundation for cost-effective amelioration of water quality problems, they must address scientific principles as well as practical ones.

Water Quality Monitoring and Assessment

Data for water quality monitoring and assessment may be divided into three types:

- data collected to characterize ambient concentrations in lakes, rivers, and ground water;
- data collected to monitor effluents; and
- data collected to monitor water quality for a specific use.

The remaining discussion focuses on ambient water quality data. However, managers of data collection programs for all three types of data need to become more aware of ways that data from individual programs can be made more useful for addressing issues that are beyond the immediate program objectives. These include:

- the need to collect important ancillary data, to place the water quality data in the context of the natural and cultural setting;
- the need to carefully document sample collection and laboratory analysis procedures; and
- the need to archive the raw data in easy-to-access computer files.

Scientific Issues and Challenges

Past experience shows that water quality data collected for utilitarian purposes are either difficult or impossible to use for scientific purposes. It is seldom appreciated that science-oriented designs not only contribute to advancing science but also significantly improve the process of attaining many practical goals.

Water quality is threatened by thousands of potentially harmful substances. Developing effective evaluations of water quality for so many chemicals is an imposing challenge, requiring continued development of screening techniques and broad-spectrum analytical procedures. We also need better ways to link contaminant selection to the physical-chemical properties of different substances, to the behavior of different substances in surface and ground water, sediments, and plant and animal tissues, to chemical usage estimates, and to the relative health and ecological risks associated with different pollutants.

A related issue has been the failure of traditional monitoring programs to identify emerging water-quality problems, possibly because of the lack of a significant link between these programs and scientific inquiry. For example, most water quality sampling in the United States has been targeted exclusively at substances for which regulations already exist, leading to a focus of effort on selected constituents—priority pollutants—that occur infrequently, and often to a disregard for more important contaminants.

Future data collection programs need to provide explicit flexibility to enable adjusting to changing environmental concerns and incorporating exploratory aspects into the design. Frequent interpretation of data also is required to identify emerging issues; the data should not be simply collected and archived for future analysis. The integration of biological measurements with physical and chemical measurements also can significantly strengthen the utility of a data collection program to help identify emerging problems. For example, biological properties may be more sensitive to water quality than are chemical or physical measurements. Too often chemical and biological measurements are considered competitive rather than complementary aspects of water quality characterization.

The design of water quality monitoring and assessment programs usually does not reflect consideration of the issue of scales. Yet the scale of focus will constrain the issues that can be addressed, for example, in providing information on non-point-source contamination. Simple use of highly intensive area sampling will not produce significant results within the limits set by realistic funding and the human resources available. Instead, innovative designs must be developed that fully use the existing understanding of the physical, chemical, and biological processes that determine water quality.

A major deficiency in environmental data collection programs has been the inadequate development of information useful for defining long-term trends in water quality. Part of the problem is simply that data collection programs are too easily abandoned when funding problems occur or in the excitement of responding to newer, more glamorous social or scientific issues. A greater commitment to continuity is needed. Moreover, a key challenge is to carefully balance long-term consistency with inevitable changes in hydrologic knowledge, the technology available for field and laboratory measurements, and the types of contaminants extant. To the extent possible, long-term programs should rely on repetition of measurement, but they must also document carefully the criteria for site selection, the characteristics of sampled sites, and the methods of data collection and analysis. When changes in measurement techniques occur, the old and new techniques should be applied in tandem as long as is necessary to determine the relationships between them.

Interrelationships among components of the hydrologic cycle must also be considered. Understanding of the connections among the atmosphere, surface water, and ground water needs to be incorporated into the design of environmental monitoring programs for these different media. For example, atmospheric cycling can be critical to the transport of major and trace constituents of terrestrial waters. So in some circumstances, a basic understanding of atmospheric processes and appropriate atmospheric monitoring may lead to more effective collection of data describing water quality.

Use of Biological Methods in Water Quality Analysis

Biological information can complement chemical analysis to improve the measurement of water quality.

Physical and chemical properties of water may vary rapidly, and intermittent or infrequent "grab" samples may give misleading indications of prevailing water quality. The native biota may be better indicators of water quality and human effects because of their prolonged exposure, integrated response, and differing sensitivity to all the varying conditions of their environment. Indeed, organisms provide the only direct measure from which ecologically significant impacts can be deduced. All levels of biological organization—molecular, cellular, tissue, organ, individual, population, and community—have been used or proposed for use in water quality interpretation. The methods may or may not identify a particular cause of change, but a measurable biological response may help to identify physical or chemical tests that should be used in the search for a cause or causes.

The first biological methods used in connection with water quality assessment were based on the observed presence or absence of species. Characteristic native species were used to demarcate zones of decreasing concentration downstream from a point of heavy organic loading. Particular species were thought to show the pollution condition in each zone. However, the supposed indicator species also occurred in unpolluted environments, and the zonation varied with the type and intensity of pollution and other hydrologic properties. Further work on human effects resulted in methods based on analysis of assemblages of species. The relative dominance of tolerant and intolerant species or of functional feeding groups in a biotic community is sensitive to water quality. These methods are successful when enough ecological knowledge exists about the species used, as is the case for most fish (although fish may be impractical to sample). They are less successful when the ecological requirements of the species are poorly known, as is usually true for algae and benthic invertebrates. In the absence of detailed information for the species of interest, effective ecological methods are available based on resemblance between biotic communities in hydrologically similar streams, with and without human impacts. The selection of suitable reference streams is crucial to the success of this approach.

The occurrence of one type of effect, sewage contamination, has traditionally been determined using as tracers microorganisms indigenous to the gut of humans and other warm-blooded animals. Bacterial density in laboratory cultures inoculated with water samples is interpreted to show the degree of fecal contamination and the potential occurrence of associated human pathogens. *Escherichia coli* is replacing fecal coliform and fecal streptococcus in these tests as a more specific indicator of human effects.

The sensitivity of organisms to target contaminants or the concentration of contaminants in living tissue can be used to detect the spatial distribution or biological availability of contaminants. The method samples native species or introduced, caged species. It is limited by differences in sensitivity or in uptake of contaminants among species, by lack of suitable widely distributed sentinel species in continental waters, and by effects of enclosure on caged organisms.

Laboratory bioassays using sensitive organisms are performed to determine biological effects of specific environmental characteristics. Responses, usually from short-term tests, are measured as bioaccumulation or as changes in behavior or physiology. Although test conditions are standardized, thus far the results cannot be extrapolated to other test conditions or species. In particular, bioassay results do not directly provide adequate information about an effect on the long-term structure and functioning of ecosystems.

Limitations of single-species bioassays have led to the use of laboratory or field-emplaced microcosms to determine the effects or the fate of contaminants. The sizes of such microcosms range from less than a liter to many cubic meters. Microcosms contain important components and exhibit important processes of natural ecosystems. They simplify environmental variability while exhibiting multispecies phenomena under controlled and replicable conditions. The results obtained from experimental microcosms are empirical analogues of whole-ecosystem functioning but require great care in broad extrapolation to the field.

Methods based on levels of organization below the individual level are applied in the field or laboratory to detect, quantify, or determine possible human effects. Techniques based on enzymes, antibodies, tissue cultures, and gene probes are being used or actively developed. The degree of sensitivity and specificity possible with these methods suggests that their use in water quality analysis will increase.

Clearly, biological data can supplement physical and chemical data to provide more holistic understanding of the functioning and of the natural evolutionary trends of hydrologic systems, as well as human effects on such systems. To accomplish this in detail, major advances are needed for determining the hydrologic implications of ecological results. Also needed are improvements aimed at increasing the sensitivity, simplicity, and uniformity of biological methods and at decreasing costs and analytical time. To date, only for indicator bacteria have procedures been adequately standardized and the results made accessible in water quality data banks. Other biological data relevant to water quality assessment are scattered and are based on diverse methods of sampling and analysis. Standardized methods would enhance the scientific value of biological information by providing a reliable baseline for making temporal and spatial comparisons. Improved communication of ecological results and their significance is also needed, in forms useful to other scientists and to the public.

Biology can furnish uncommon insights for hydrologic science, insights not achievable solely from a knowledge of physics and chemistry. For example, organisms are involved in the transport and cycling of elements in water and sediments. Organisms are targets of scientific efforts to preserve rare and endangered species. Populations of organisms are intentionally affected by management programs and unintentionally affected by natural and anthropogenic environmental effects. These and other issues often require studies on large spatial and temporal scales. Such studies should be incorporated into national and international water quality monitoring systems to provide the means to evaluate and improve incompletely developed but potentially valuable biological methods for understanding the organization and functioning of hydrologic systems.

SOURCES AND SUGGESTED READING

Andre, J. C., J. P. Goutorbe, and A. Perrier. 1986. HAPEX-MOBILHY: A hydrologic atmospheric experiment for the study of water budget and evaporation flux at the climatic scale. Bull. Am. Meteorol. Soc. 67:138-144.

Baumgartner, A., and E. Reichel. 1975. The World Water Balance. Elsevier, Amsterdam, 179 pp.

Bernabo, C. 1978. Proxy Data: Nature's Records of Past Climates. NOAA Environmental Data Service Report. U.S. Department of Commerce, Washington, D.C.

Earth System Sciences Committee, NASA Advisory Council. 1988. Earth System Science: A Closer View. National Aeronautics and Space Administration, Washington, D.C.

EOS Science Steering Committee. 1987. Earth Observing System. Vol. II. From Pattern to Processes: The Strategy of the Earth Observing System. National Aeronautics and Space Administration, Washington, D.C., 140 pp.

Haeni, P. 1983. Sediment deposition in the Columbia and Lower Cowlitz rivers, Washington-Oregon, caused by the May 18, 1980, eruption of Mount St. Helens. U.S. Geological Survey Circular 850-K, 21 pp.

Kinter, J. L., and J. Shukla. 1990. The global hydrologic and energy cycles: Suggestions for studies in the pre-global energy and water cycle experiment (GEWEX) period. Bull. Am. Meteorol. Soc. 71(2):181-189.

Krishnaswami, S., W. C. Graustein, J. F. Dowd, and K. K. Turekian. 1982. Radium, thorium and radioactive lead isotopes in groundwaters: Application to the in situ determination of adsorption-desorption rate constants and retardation factors. Water Resour. Res. 18(6):1663-1675.

Robertson, W. D., and J. A. Cherry. 1989. Tritium as an indicator of recharge and dispersion in a ground water system in central Ontario. Water Resour. Res. 25:1097-1109.

Ryan, P. F., G. M. Hornberger, B. J. Cosby, J. N. Galloway, J. R. Webb, and E. B. Rastetter. 1989. Changes in the chemical composition of stream water in two catchments in the Shenandoah National Park, Virginia, in response to atmospheric deposition of sulfur. Water Resour. Res. 25:2091-2099.

Skinner, B. J., and S. C. Porter. 1989. Physical Geology. John Wiley & Sons, New York.

Smith, R. A., and R. B. Alexander. 1986. Correlations between stream sulphate and regional SO_2 emissions. Nature 322:722-724.

Stockton, C. W., and G. J. Jacoby. 1976. Long-term surface water supply and streamflow trends in the upper Colorado River basin. Lake Powell Research Project Bulletin Number 18. University of California, Los Angeles.

U.S. Committee for an International Geosphere-Biosphere Program. 1986. Global Change in the Geosphere-Biosphere: Initial Priorities for an IGBP. National Academy Press, Washington, D.C.

Education in the Hydrologic Sciences

Higher education in hydrology, especially at the graduate level, has long been the province of engineering departments in most universities. Doctoral and master's degree programs administered by these departments have been directed toward the traditional concerns of water resources development, hazard mitigation, and water management as predicated on societal needs. The research focus in these departments has properly been the analysis and solution of problems related to engineering practice, on the premise that these problems contribute palpably to the technical knowledge base required for water resources allocation, the management of floods and droughts, and pollution control. Current societal needs, as expressed through legislative action or executive orders, are as important to the choice of research problems and their methods of solution as are the flow of scientific ideas and technological breakthroughs.

This well-developed and successful line of inquiry differs markedly from that pursued in the pure sciences, such as chemistry. The difference, in fact, is exactly analogous to that between the disciplines of chemistry and chemical engineering. Chemistry is the science that deals with the composition, structure, and properties of substances and the reactions that they undergo. Chemical engineering deals with the design, development, and application of manufacturing processes in which materials undergo changes in their properties. The first discipline is a science, dealing with puzzle solving (i.e., motivated by a question), whereas the second is an application of science, dealing with problem solving (i.e., motivated by the solution). Hydrology has a long and distinguished history of problem solving, but where is the antecedent science of puzzle solving?

The education of hydrologic scientists offers challenges as great as those in engineering hydrology, but the spirit of the enterprise is different, just as it is between education in chemistry and in chemical engineering. In scientific hydrology, as in chemistry, research is done in the context of the three chief stages of development of any pure science: careful observation of phenomena (the natural history stage), quantification and conceptual modeling (the empirical stage), and quantitative prediction (the exact stage). The choice of research problem is occasioned by its level of development within the hierarchy of the science, by the availability of new methods with which to solve it, and by the desire to understand a hydrologic phenomenon more deeply. The solution of the problem advances the development of the science and expands the conceptual framework that gives it meaning. It is this kind of internally driven intellectual pursuit that motivates the pure scientist and that must be instilled by the educational process that forms her or his professional outlook. That is the challenge to hydrologic science, and it differs from the challenge to engineering. It is a challenge that must be met at the graduate and undergraduate education levels, in precollege education, and in educating and training an increasingly diverse student population.

GRADUATE EDUCATION IN THE HYDROLOGIC SCIENCES

As a result of this challenge, graduate education in the hydrologic sciences should be pursued independently of civil engineering. The problem is made clear by the disciplinary structure of earth system science as illustrated in Figure 5.1. The warp of this intellectual fabric consists of the three traditional geoscience threads: solid earth science, atmospheric science, and ocean science. The weft contains multidisciplinary threads among which hydrologic science is dominant by virtue of its central role in cycling energy and matter. Some universities have recognized this by housing "water science" programs in departments such as geography or geology. However, few offer a coherent program that treats hydrology as a separate geoscience. It is a premise of this report that hydrology—expanded in scope, importance, and potential—must escape mere inclusion as an option under engineering, geology, or natural resources programs. Establishment of specialized Ph.D. and master's degree programs is, therefore, necessary to enhance the identity of hydrology as an established science. Graduates are needed who are considered first and foremost as hydrologists, not as civil engineers or geologists who know something about hy-

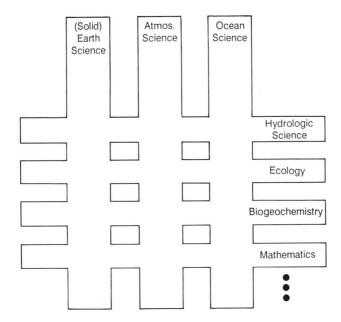

FIGURE 5.1 The disciplinary structure of earth system science.

drology. A solid program of course work with unified requirements would constitute an integral part of a graduate program and thereby ensure that degree candidates in the hydrologic sciences have a common background in fundamental, scientific hydrology.

Besides these professional considerations, there are institutional constraints that lead to the conclusion that **a hydrologic sciences program should not be "hosted" by a single department in another discipline.** Consider, for example, the case in which hydrology is viewed as a subdiscipline within a geological sciences department. In the more traditional geology departments, students encounter serious difficulties in preparing for comprehensive examinations because of departmental policies that impose a geological focus on these examinations. Geologists, like other disciplinary scientists, tend to be conservative when defining requirements pertaining to the main elements of their subject. These narrow requirements are appropriate for students in, say, petrology, geochemistry, or petroleum geology, but they do not serve students well who specialize in the hydrologic sciences. The committee for a Ph.D. comprehensive examination is a departmental committee, usually with limited flexibility in crossing departmental boundaries; often this is also true of the research committee. A similar problem can occur in civil engineering departments, where graduate students who do not have an undergraduate degree in engineering may be required to complete a suite of core courses in the undergraduate engineering curriculum, irrespective of issues related to professional accreditation and to the detriment of a specialization in hydrologic science.

There are three potential options for structuring a graduate hydrology program that is not a subdiscipline in a host geological sciences, geography, or engineering department. One option is a separate department of hydrologic science. The other two options involve less formal, multidisciplinary programs, in one case autonomously degree-granting and in the other, degree-granting through participating departments. Each option offers advantages and disadvantages with respect to the needs of university administration, faculty, and students.

The first option is perhaps the ideal one: establishing a graduate department. It is probably also the least realizable in most universities, at least in the near future, given the usual resistance to the creation of new academic units with autonomous status and dedicated facilities. Nonetheless this approach best serves the goal of establishing hydrology as an integral geosciences discipline, with a distinct identity separate from engineering. This option also avoids certain pitfalls of multidisciplinary programs, which are described below. Very few universities, however, will have the commitment and resource base necessary for introducing a new department. In most cases, such an academic unit would probably be limited to a few core faculty members with adjunct professors from other departments. Creation of a separate department is a goal to strive for, however, at a few key universities where current, well-established hydrology programs make this option viable. An additional resource base might be available with federal funding for a center of excellence in hydrology, or some comparable concept. In most universities, however, multidisciplinary programs are the more feasible and realistic approach.

A multidisciplinary, interdepartmental program has some unique advantages over a separate department. Hydrologic science is, by its very nature, interdisciplinary (see Figure 5.1) and hence is well suited to such a format. The courses taught presently in hydrology at academic institutions show its diversity, since they are offered typically in a range of departments and programs (e.g., civil engineering, forestry, geology, geography, and soil science). Moreover, faculty members with strong interests in hydrology, although they may not teach more than one course in the area, are also found in a diverse array of disciplines (e.g., aquatic ecology, limnology, and meteorology).

Adding to this interdepartmental flavor is the breadth expected of doctoral candidates who wish to do research in the hydrologic sciences.

For example, if research focuses on global climate and hydrology, the perspective to be developed in a student is that of the geophysicist who elucidates the dynamical, thermal, and hydrologic interactions among the atmosphere, ocean, and land surface. This comprehensive effort involves theoretical analysis, numerical modeling, laboratory experiments, and the analysis of observation. For research on ground water hydrology, the required perspective is a quantitative focus on the analysis of ground water flow and mass transfer, with an understanding of the fundamental role that geology plays in determining the nature of the subsurface environment. And for research on the chemistry of hydrologic processes, the perspective to be developed in the student is that of the applied chemist who is comfortable working with field scientists (or with engineers) and who appreciates the relationship between pure chemical research and the behavior of open, natural water systems. These three examples illustrate the need for a broad range of educational inputs to graduate education in the hydrologic sciences.

But the interdepartmental option has some disadvantages. One is that students often fail to achieve a disciplinary perspective or a feeling of attachment to a professional discipline. Moreover, many of the courses, when taken from a set of departmental rosters, are tailored to the needs of the departments and not to the field of hydrology. The hydrology graduate student may be inadvertently short-changed. From the faculty perspective, allegiance is split between the program and the home department, two units whose goals are not necessarily compatible. The multidisciplinary approach also creates an additional layer of administration. Nevertheless, this is probably the most appropriate approach for many academic institutions.

For a multidisciplinary, interdepartmental program, both degree-granting and non-degree-granting options must be considered. The degree-granting option is advantageous for the student. It minimizes the problems of satisfying department course and comprehensive examination requirements that may not be closely relevant to hydrology studies. It also provides a higher probability of receiving financial resources (e.g., research and teaching assistantships) than when these are linked to member departments whose first priority in granting financial aid is to support their own students. This is, of course, also true for resources such as laboratory and office space, or equipment.

Although the degree-granting option may be preferable from the students' perspective, it is not practical to envision the creation of a degree-granting academic unit in the hydrologic sciences on most campuses. Interdepartmental graduate programs that grant degrees are confederations that must compete with degree programs in the member departments for resources, students, and faculty loyalties. Usually this competition will not turn out favorably for the multidisciplinary program, since it has neither the professional strength nor the tradition of faculty support that has been garnered by the disciplinary programs. The onus of attempting to overcome these obstacles to survival is lessened if a graduate program in the hydrologic sciences offers degrees only through member departments.

The advantages of a multidisciplinary, non-degree-granting graduate program are in reducing conflict, both between the program and the participating departments and among the participating departments, over admission requirements, resource allocation, faculty time and effort, facilities, and extramural support. These issues would not generate controversy because they would be treated by each department individually, with concurrence required only in respect to the general design of the interdisciplinary graduate program to which all departments have contributed. The disadvantages of the program not granting a degree are in the loss of direct visibility as an academic unit, an increase in the bureaucracy attending academic planning, and a reduced opportunity for faculty interaction in research and teaching. These disadvantages must be weighed against the benefits from lessened conflicts before deciding how the hydrologic sciences program should be structured.

STRUCTURING THE GRADUATE PROGRAM

A solid program of course work with unified requirements would constitute an integral part of any graduate program in the hydrologic sciences and thereby ensure that degree candidates would have a common background in fundamental scientific hydrology. The course program would introduce students to a broad range of hydrologic processes and would form the basis for further specialization in surface water or ground water hydrology, global climatic processes, hydrometeorology, hydrogeochemistry, or surficial processes. This formal core curriculum should be rounded off with multidisciplinary seminars addressing issues related to environmental quality and including scientists, engineers, economists, and water managers.

Table 5.1 lists four general areas of course work that can serve as the basis for a core curriculum in the hydrologic sciences at the graduate level. Each topic entered under one of the four areas can be the subject of a single course or can be included with other topics in a single course—the precise structuring of the curriculum will vary among programs. In most cases, a field course that integrates the contents of the classroom courses may be desirable in order to obtain

TABLE 5.1 A Set of Topics for Graduate Programs in
the Hydrologic Sciences

General Areas	Individual Topics
Fluid motions	Flow in porous media Geophysical fluid mechanics Open-channel flows Theoretical or dynamic meteorology
Hydrologic phenomena	Aquatic biology and ecology Aquatic chemistry Boundary-layer meteorology Climatology Fluvial geomorphology Geochemistry Ground water hydrology Hillslope hydrology Microbiology Soil physics Snow hydrology Surface water hydrology
Hydrologic techniques	Computer simulation Data analysis methods Field research methods Optimization and decision analysis Remote sensing Software development Statistical inference Stochastic processes
Hydrologic policy	Natural resource economics Water law and institutions Water resource management Water quality management

a more direct appreciation of hydrologic processes than can be had
from classroom materials, laboratory exercises, and weekend field
trips. A field course could be designed to allow students to develop
a better understanding of runoff generation processes, survey different hydrologic regimes, illustrate the water infrastructure of a state,
or introduce and demonstrate methods of field data collection.

The first step in establishing a graduate program in the hydrologic
sciences should be the convening of interested faculty and administrators
on a campus to form a working group that will develop an academic
plan. This plan should consider such elements as:

VEN TE CHOW
(1919-1981)

Ven Te Chow was a premier synthesizer of hydrologic knowledge. His books *Open-Channel Hydraulics* and the *Handbook of Applied Hydrology* remain classics in the field and are still in print more than a quarter century after they were first published. Over that period, the hydraulics book has sold more than 40,000 copies and the handbook 25,000 copies. Perhaps more than any other book, the *Handbook of Applied Hydrology* defined hydrology as a subject, and it has stood as a basic hydrologic reference for many years.

Ven Te Chow was born and raised in China. After being educated in Shanghai, he taught civil engineering there during the war years. His literary skills and determination to succeed were apparent even at that time, and by the age of 24 he had already published his first book, in Chinese, on the theory of structures. He came to the United States in 1947, first for a master's degree at Pennsylvania State University, and then for doctoral studies at the University of Illinois.

Thus began what was to become a life-long association with the University of Illinois, where he served on the faculty for more than 30 years. It is remarkable that although he did not specialize in water studies before coming to the United States, he wrote his first landmark book within a decade of arrival in this country.

Besides being an outstanding author, he was also an innovative researcher, pioneering in 1951 the frequency factor method, now the basis of extreme flood computation. He did much to foster the concept of systems modeling in hydrology and coined the term "hydrosystems engineering" to refer to the interaction between hydrology and other technical, social, or economic factors.

Besides making academic contributions, he was prominent in professional organizations and was instrumental in founding the International Water Resources Association, a worldwide organization serving all aspects of water resources activity. He was its first president, from 1972 to 1979. He was also president of the Hydrology Section of the American Geophysical Union from 1974 to 1976.

His books reflect the logical mind that was the hallmark of his organization of knowledge. He had an almost unique ability to plumb the depths of theoretical research and also express the results in a form that the average hydrologist could understand. His wife once asked him what he thought was the secret of his success, and he responded that perhaps he was more patient than others. Certainly his inexhaustible attention to detail was remarkable.

And beyond his scholarly capabilities, Ven Te Chow possessed a rare feeling for people, a fundamental understanding and appreciation for human dignity that underlay all that he did. *Open-Channel Hydraulics* is dedicated to "Humanity and human welfare." He was able to accomplish his purposes with the lightest of touches, always leaving others with an alternative path should they wish to choose it.

- the institution's current effort to provide graduate education in hydrology, focusing on its strengths and weaknesses as a coherent program;
  - graduate hydrology programs at comparable universities;
  - the current and potential capability of the present faculty to address critical and emerging areas of hydrologic research such as are specified in this report;
  - programmatic areas in the hydrologic sciences that are essential but not addressed currently;
  - the potential for cooperation in research and teaching among the host departments;
  - administration of the graduate program as a department, inter-departmental degree-granting program, or non-degree-granting program;
  - faculty recruitment needs;
  - facilities and space needs (laboratory and field);
  - technical staff support needs;
  - admission requirements for Ph.D. and M.S. programs;
  - degree requirements; and
  - student recruitment.

Once an academic plan is established, procedures can be developed to implement the plan. Typically, once a program has begun, a campaign to recruit students commences. Prospective students would apply for admission to an appropriate academic unit. After entering, the students would be assigned a major adviser with whom to consult about a specific academic program as soon as possible to secure initial approval of the program. Graduate students would be expected to take courses in several disciplinary areas, but these courses would have to conform to the requirements of the hydrologic sciences program. Flexible academic curricula should be developed to enable graduates from the pure sciences and other fields to obtain graduate degrees in the hydrologic sciences without an excessive number of remedial courses.

Because of the multidisciplinary nature of the hydrologic sciences, students from widely different backgrounds are likely to be attracted to the discipline. Some will come from the basic sciences because they find the analytical complexity of hydrologic problems exciting. Others will have a background in other environmental sciences, but without substantial preparation in mathematics. Still others will have worked as field-oriented scientists or on field projects. The graduate program must recognize this diversity. Outlined below is a list of components of an undergraduate-level preparation for study at the graduate level, but it is recognized that some potentially excellent students will not have completed all of the following requirements:

- substantial background in one of the earth, life, or atmospheric sciences, e.g., biology, forestry, geography, geology, meteorology, and soil science;
- courses in the supporting pure sciences, e.g., physics and chemistry;
- mathematics through differential equations, linear algebra, statistics, and probability theory;
- experience with measurement of natural phenomena, preferably in field situations as well as in controlled laboratory settings;
- familiarity with computers, including programming in higher-level languages, mathematics and statistics software packages, graphics, and text processing; and
- experience in writing short research papers, based not only on familiarity with published papers, but also requiring analysis of data.

Extramural support for a new graduate program in the hydrologic sciences is essential, in the form not only of research grants but also of research fellowships. Ideally, the National Science Foundation and other funding organizations could institute a program of predoctoral fellowships with an emphasis in hydrology. Such a program would attract students to the hydrologic sciences and train them specifically; impart a degree of autonomy to the graduate program at the Ph.D. level; and represent money spent efficiently, without overhead costs but with a focused objective. Perhaps most important, it would help to build a national base of highly trained, multidisciplinary scientists in a critical area of significant potential impact on society.

UNDERGRADUATE EDUCATION IN THE HYDROLOGIC SCIENCES

Few undergraduate programs exist in hydrology, and most professionals gain entry to the field from engineering or from the geosciences. This point is illustrated in Figure 5.2, which shows the distribution of academic backgrounds of hydrologists employed by the Water Resources Division of the U.S. Geological Survey in 1986, when the division had 2,055 professional employees. Of these, 85 percent held the title of hydrologist. Figure 5.2 shows that about half of the Water Resources Division's professionals classified as hydrologists had majored in geology, civil engineering, and environmental (or sanitary) engineering. Not shown are trends in this background with time, but it is probable that the geology and civil engineering portion has decreased significantly during the last decade or two.

The existence of an undergraduate population prepared for and

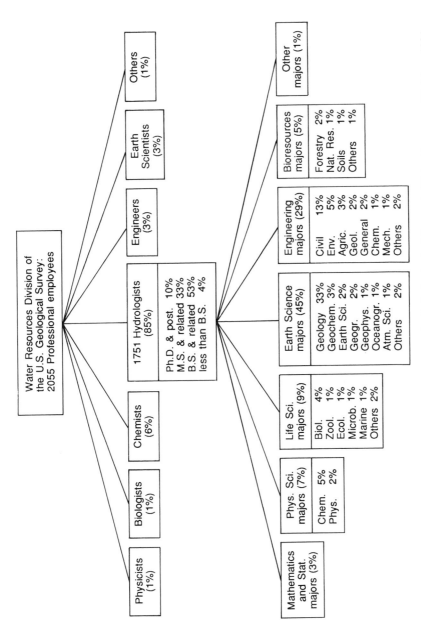

FIGURE 5.2 Academic backgrounds of hydrologists in the U.S. Geological Survey's Water Resources Division, 1986.

interested in graduate work in hydrologic science depends on what happens to potential hydrologists during their undergraduate studies. Indeed, the geosciences and civil engineering have suffered a precipitous decline in undergraduate enrollment in recent years. The number of majors declined by two-thirds from 1982 to 1987, while graduate degrees remained fairly constant, with some increase in the late 1980s. The number of undergraduate degrees in civil engineering fell by 27 percent between 1983 and 1989. This sharp decline in undergraduate degrees in the traditional contributing disciplines is beginning to have an impact at the graduate level, thus affecting the overall health of the hydrologic sciences.

Thus the hydrologic sciences face a potential recruitment problem created, at least in part, by new, rather general trends among young people that reduce the number aspiring to a scientific career. Such trends intensify the competition for students, and the hydrologic sciences must face this fact. The recruitment problem may become especially acute because of increasing demands and opportunities that will require an increase rather than a decrease in the number of hydrologic researchers.

The task of enlarging the pool of undergraduate students in the hydrologic sciences may be hindered more by students' inadequate mathematics and science background than by lack of interest. As a case in point, a survey by the American Geological Institute showed that, of the 340,000 freshmen planning for degrees in the natural sciences and engineering in 1980, only 206,000 were degree recipients in these areas four years later. This attrition is at least partly explained by the increasing difficulties students face as they enroll in courses in these majors, the primary obstacle being the required capabilities in physical science and mathematics. This obstacle, in turn, results from a de-emphasis in these areas in precollege education and the failure of universities to establish mathematics and science entrance requirements. The students who could master such courses, were they exposed to them earlier, either lack the motivation to enroll voluntarily in science and mathematics or are convinced psychologically that such courses are beyond their abilities. The solutions to this problem lie in an enhanced science and mathematics curriculum at the precollege level, with encouragement for students who did not acquire this background to believe that college-level courses in mathematics, chemistry, and physics are within their abilities.

New demands and opportunities will also create an ever-increasing call for data collectors, laboratory analysts, technicians, and field assistants. In particular, the need for spatially broad, detailed, and sophisticated data collection systems, already considerable, may greatly expand with intense national interest in water quality and climatic change. The personnel needed for such activities are hydrologists with degrees below the Ph.D. level. The quality of their preparation will determine the quality of data generated for both applied and scientific purposes. (It should be noted that undergraduate studies determine this preparation for B.S.-level and, to a large degree, for M.S.-level hydrologists.)

This fact brings up another serious educational problem—**the lack of field and laboratory experience at the undergraduate level, a situation that has reached crisis proportions**. The almost complete disappearance of laboratory education can be attributed to many factors, most of them related to funding. Laboratory courses demand both facilities and high faculty-to-student ratios, but many universities lack the resources to finance facilities and the teaching assistants, technicians, and machinists required to support them. The philosophical framework of science education also has changed, emphasizing intellectual instead of technical skills. Moreover, faculty members whose expertise and effort are centered on field or laboratory experimentation, instrumentation, and technical methods are perhaps at a disadvantage when considered for advancement because they usually publish fewer papers.

Finally, the nearly universal demand for computer literacy has left students with little time for commitment to laboratory and field courses. This is a problem at all levels that has existed long enough to become self-perpetuating through the next generation of faculty. The consequences of this deficiency are both profound and disturbing. Students have become separated from the realities of the physical world they seek to master, studying only conceptual models in which the rich complexity of nature is replaced necessarily by the convenience of ad hoc simplification. In the absence of experimental validation, these models tend to take on an aura of reality in the minds of their users, which may lead to scientific error and stagnation. If a major rejuvenation of the "observational" components of higher education were to occur, it would serve to improve the quality of professionals entering hydrologic science and also perhaps to attract larger numbers of experientially motivated students to the field.

In spite of the importance of the above needs, the role played by undergraduate studies in the hydrologic sciences often is underestimated. Perhaps this happens, in the United States and elsewhere, because of the almost complete absence of academic departments devoted to scientific hydrology. There is no department responsible for basic requirements. However, the increasing urgency of the needs related to the hydrologic sciences calls for a special effort to satisfy them. Obviously, one way to do that is to create, at enough universities, the appropriate department of hydrologic sciences that includes undergraduate majors. However, as discussed above, such a task is difficult and probably unachievable; at best, its aims would be implemented too slowly to satisfy near-future needs. Another approach is to increase the duration of studies leading to M.S. and Ph.D. degrees in the hydrologic sciences. However, such an action might make graduate studies less attractive and, therefore, be counterproductive. A more reasonable course of action is to include activities designed to influence undergraduates as a major part of the programs administered by interdepartmental committees (or groups) assisting with and organizing graduate work in the hydrologic sciences.

The undergraduate experience in science and engineering can be modified in several ways to facilitate the education of hydrologic scientists who will emerge from these disciplines and to promote multidisciplinary science. For example, faculty in related disciplines can enhance awareness of hydrology and guide students to graduate programs or professional positions in hydrology. Another possibility is increased acceptance of engineering-based hydrology courses as electives in liberal arts and science degree-granting programs. The few existing hydrology departments also can help by adding to their program one or two courses from other disciplines or by revising water-related courses in ways that will attract students from other sciences. Some specific activities designed to influence undergraduates could include (1) development of a course list outlining undergraduate preparation for a career in scientific hydrology; (2) dissemination of such a list among undergraduate science and engineering advisors (and an attempt to get them to use the list); (3) organization of a solid (perhaps senior-level) course in scientific hydrology; (4) sponsoring, at the university level, periodic public lectures about some of the interesting problems investigated in scientific hydrology; and (5) initiating and administering a multidisciplinary undergraduate major in scientific hydrology.

SCIENCE EDUCATION FROM KINDERGARTEN THROUGH HIGH SCHOOL

The discussion above makes clear that the success of graduate programs in the hydrologic sciences will depend on the quality of undergraduate preparation in pure science and mathematics, which, in turn, depends critically on the educational background obtained in precollegiate years. Like the statistics quoted above for geosciences and civil engineering majors, those for science education among high school students show a dismal trend.

Less than 50 percent of high school graduates in the United States have completed more than one year of mathematics and one year of basic science. Less than 10 percent have taken a physics course. Students in Europe, the USSR, and Japan take considerably more mathematics and science courses than do North Americans.

This decline in high-quality, well-attended science programs has been a concern for many years to educators and leaders in business and industry. Recognition of the need for stronger science education programs has led to a reexamination of curricula in primary and secondary education in all states during the past decade. Parents and educators are showing a renewed interest in science, which comes at a time when student interest also is growing. Student enthusiasm for science-oriented programs has never been higher than it is today, and the demand for new technical employees in science, engineering, and computer-assisted technology continues to accelerate.

Intensive, high-quality learning experiences in science in preparation for college require support from parents willing to continue their own education; from schools willing to upgrade their curricula and expand the diversity of science courses; from universities willing to assist in teacher education and staff development; and from businesses and research agencies willing to share their technical expertise and equipment with the schools. Program improvement can be achieved gradually by changes in school district policies or state laws and more rapidly by school-level planners utilizing, on an ad hoc basis, a diverse spectrum of opportunities to improve science education.

Early in the 1980s, a California study of the attitudes and education of teachers regarding science education provided the following information:

• Only 5 percent of California school districts employ full-time science specialists.

• Student participation in science activities averaged 44 minutes per week in elementary schools.

• Over 40 percent of the elementary school teachers surveyed rated their own ability in science as below average when compared with other subjects. The felt they did not have the skills to teach science processes and concepts.

• Forty-five percent of the elementary school teachers and administrators predicted that less money for instructional materials will be spent on science because state funds provided for instructional materials may be spent on any subject area districts or schools see fit within established guidelines.

• Although teachers expressed their support for the "hands-on" concept, most continue to use textbooks (56 percent) or teacher-made written materials (57 percent) as the basis for their science instructional programs.

• Although science kits or systems have been purchased by many schools or districts, only 5 percent of the teachers use them extensively.

• Many teachers expressed the feeling that other subjects took priority in time over science.

Given these realities at the elementary school level, what is the probability of seeing a high-quality science background developed in students at the high school level? It is obvious that staff development for both teachers and administrators will play a pivotal role in the improvement of science education. Hydrologists have a challenge and an excellent opportunity to influence and accelerate that development.

WOMEN AND ETHNIC MINORITIES IN THE HYDROLOGIC SCIENCES

That the United States faces a shortage of technically trained personnel in the next decade and beyond is a problem well recognized in the scientific and engineering communities (Widnall, 1988; TFWM, 1989). Traditionally these fields have been dominated by white males. As the rate of white males entering these professions continues to decline, however, attention has turned to those populations who have been underrepresented and underutilized in science and engineering—women and minorities. If the nation is to be able to meet its needs for scientific personnel into the next century, greater numbers of women and minorities will have to be recruited and retained in science and engineering professions (TFWM, 1989; NRC, 1989b).

The hydrologic sciences face perhaps an even greater challenge in meeting national needs for technically trained personnel than do other scientific disciplines. As the demand for master's-level hydrologists in government and industry increases, individuals who may have otherwise pursued a Ph.D. often opt to enter the work force. Universities find it difficult to compete with the high salaries and hands-on experience offered by consulting firms. Thus it is necessary not only to recruit more young people to fill the ranks of the hydrologic sciences as the traditional source of students dwindles, but also to recruit and train master's-level hydrologists to meet the nation's growing need for these skills.

While the challenge is to improve recruitment and retention of persons of all types of background, women and racial and ethnic minorities face certain obstacles that require particular attention. The underrepresentation of these populations in science and engineering is well documented. The growth in the employment rate for women scientists more than doubled the rate for men between 1976 and 1986 but has slowed in recent years. In 1986, while women accounted for 44 percent of the U.S. work force, they accounted for only 27 percent of all scientists (including social scientists) and engineers. The number of women planning careers in science or engineering peaked in the late 1970s and is now declining. While blacks constitute 12 percent of the general population, only 2 percent of all employed scientists and engineers are black. Hispanics, America's fastest growing minority, account for 9 percent of the population but only 2 percent of all employed scientists and engineers (TFWM, 1989).

The pattern of underrepresentation is mirrored in the numbers of graduate degrees earned by women and minorities. In 1987, women received 16.7 percent of all doctorates in the physical sciences and 6.5 percent in engineering, according to the National Research Council's Doctorate Records File (DRF), which includes data on individuals receiving Ph.D.s from U.S. universities (NRC, 1989a). This picture is only a little less discouraging than that for minorities achieving advanced degrees. From 1985 to 1987, the number of American ethnic minorities earning science and engineering B.S. degrees rose 21.6 percent. At the master's level, the increase was 9 percent, and at the Ph.D. level the number of degrees awarded remained static over the two-year period (Vetter, 1989). About 1 percent of the doctorates awarded in science and engineering are earned by black Americans, and about 2 percent are earned by Hispanics (TFWM, 1989). While in 1977 blacks earned 684 science and engineering doctorates, in 1988 that number had fallen to 311 (Vetter, 1989).

The hydrologic sciences follow the national trend of underrepresentation of women and minorities in their ranks. This committee's 1988 survey of the backgrounds of hydrologists demonstrates the case with respect to women (see Appendix B). Of the 2,200 persons who responded to the survey, 11 percent were female and 89 percent were male. Of the males, 54 percent held Ph.D. degrees, while only 28 percent of the women did. Data from the DRF's 1987 survey of earned doctorates (NRC, 1989a) indicate that of the 18 Ph.D.s earned in hydrology and water resources, 11 were earned by U.S. citizens and individuals with permanent visas; of these 11, 5 were awarded to women and none to minorities. In 1988 a total of 24 doctorates were awarded in hydrology and water resources, of which 14 were awarded to U.S. citizens or individuals having permanent visas (NRC, 1989a). Of the 14 recipients, 13 were white, 1 was Hispanic, and 4 were women.

The reasons that women and minorities traditionally have not chosen careers in science and engineering are diverse—some are clear, while others are subtle and not well understood. Sex-related inequalities in measures of career success, such as academic rank, tenure, and salary, may influence young women to seek careers in other professional fields where they perceive less inequality. Women in science and engineering fields have lower rates of recruitment and retention than do men. After expressing an initial interest in science or engineering studies, women, more often than men, switch to nonscience or nonengineering fields. Typically such decisions are based on sociocultural or attitudinal factors rather than academic talent. Although the rate of attainmnent of Ph.D.s remains lower for women than for men in most fields of science and engineering, there is no indication that this attrition is due to a lack of academic performance. Research indicates that, especially for women, the attrition from science and engineering majors is seldom related only to academic talent and achievement (NRC, 1989b).

Women's decisions to marry and have families often are said to lead to career decisions that benefit their families but damage their careers, decisions that males are seldom, if ever, forced to make. Women scientists and engineers more frequently attribute part-time employment and time spent unemployed and not seeking work to family obligations; also, married and single female academics are less geographically mobile than their male counterparts, which may hinder their potential for career advancement. However, data show that marriage and parenthood do not result in lower publication rates or lower rank and salary among women (NRC, 1987). Although the unique demands of marriage and parental responsibilities on women are not the sole cause of low rates of recruitment, retention, and success of female academics, they are likely contributors. To reduce the loss of women in engineering and the sciences, academia must develop more flexibility to accommodate the unique set of demands faced by women.

Other, more subtle explanations for the lack of women in the sciences and engineering, particularly in academia, have been reported. Women graduate students often believe they are subject to inappropriate treatment by male faculty and student colleagues. "Inappropriate treatment" is defined as "any treatment that emphasizes the student as a woman first and a student second and stresses the social nature of an interaction instead of the professional or educational nature." Widnall (1988) reports that there are still male faculty members in science and engineering who state publicly that women do not belong in graduate school. Zikmund (1988) states that the negative experiences of women on faculties and in college administrations are not random and that the well-being of academic women is still being sabotaged in subtle ways. A large percentage of women responding to a Massachusetts Institute of Technology survey believed their gender was a significant barrier to receiving academic resources. The current environment that women face in graduate school thus may be a major reason for the small number of women today in science and engineering programs. An increased willingness on the part of faculty to challenge professional colleagues who make prejudicial or inappropriate remarks about women and minority students could help to reduce these types of negative experiences.

The challenge for the sciences is to increase the number and diversity of students at all educational levels. Although successful approaches to this problem will necessarily be as diverse as the disciplines, institutions in which they are housed, and individuals who pursue them, there are some fundamental principles that have been demonstrated to work. In its 1989 report, *Changing America: The New Face of Science and Engineering*, the congressionally mandated Task Force on Women, Minorities, and the Handicapped in Science and Technology lists actions to be taken to increase the participation of underrepresented populations in science and engineering, and identifies exemplary programs of this nature (TFWM, 1989).

Experts in the field believe that increasing the participation of underrepresented populations should be approached as a systems problem, requiring coordinated changes in policies and procedures rather than isolated, radical interventions without continuity. Effective programs must be implemented in all sectors of education, beginning at the prekindergarten level and continuing through employment (NRC, 1989b). At the precollege level, emphasis on the usefulness of mathematics and science training has been shown to make these subjects more attractive to female students (NRC, 1984). Expectations should be raised for all students, with hands-on science education provided in a forum free of cultural and gender biases. Counselors should encourage all students to consider science and engineering careers and emphasize the importance of mathematics and science proficiency in the job market of the future (TFWM, 1989).

At the college and graduate levels, educational institutions should improve recruitment and retention programs. The use of role models and mentors in education has been demonstrated to be effective in recruitment and retention programs (TFWM, 1989; NRC, 1989b). Currently, however, female and minority faculty members are most likely to be found among the untenured junior faculty and thus not available for significant amounts of time to serve as mentors. Recruitment

THE CHANGING PROFILE OF THE HYDROLOGIC COMMUNITY

In 1962, the Federal Council for Science and Technology published the results of a survey of the 811 individuals self-listed with the National Register of Scientific Personnel as hydrologists. Twenty-six years later, a similar but more detailed survey was conducted as a part of this study among the approximately 3,000 members of the Hydrology Section of the American Geophysical Union. A remarkable 2,200 responses were received from this latter survey. Following are some comparative highlights from the 1960 and 1988 surveys, whose results are reported in detail in Appendix B of this report.

- The average age of hydrologists has not changed significantly: 43 years in 1960 and 42 years in 1988. Among female hydrologists, however, the average age was 35 years in 1988.
- In 1960, most hydrologists in the survey held bachelor's degrees (74 percent), whereas in 1988 most held Ph.D. degrees (51 percent), with the number of bachelor's degree holders dropping to 11 percent. (This trend may reflect to some extent a greater propensity among doctoral degree holders to join the American Geophysical Union than among bachelor's degree holders.)
- In 1960, engineering (55 percent) and geology (28 percent) were the most important degree fields for hydrologists. In 1988, the situation was unchanged with respect to geology (still 28 percent), but engineering had dropped to 35 percent, while hydrology had risen to 14 percent as a degree field. Overall, 77 percent of the hydrologists responding in 1988 were trained as engineers, geologists, or hydrologists.
- About two-thirds of the hydrologists responding in 1960 worked for the federal government, whereas less than one-third did so in 1988. Private sector employment rose from 10 percent in 1960 to 32 percent in 1988. Most of those in the latter group were working in ground water hydrology. In 1988, the federal government, educational institutions, and the private sector each accounted for about one-third of the total employment of hydrologists. Forty percent of all hydrologists are employed in applied hydrology today.
- The ratio of surface water to ground water hydrologists was about 2:1 in 1960. In 1988 the ratio was about 0.7:1.
- Most hydrologists with graduate-level degrees choose specialties in the earth's crust and landforms or chemical processes.

These results suggest that over the past three decades hydrologists have become more highly educated, more likely to have a geosciences background, more evenly distributed among the public and private sectors of employment, and more involved with ground water phenomena, including chemical quality. Female hydrologists tend to be younger, less highly educated, and more geoscience oriented (as opposed to engineering oriented) than their male colleagues.

and retention of underrepresented students in science and engineering departments would be further enhanced by the presence of women and minorities at all ranks, a signal to such students that they would be respected and treated fairly.

The availability of financial support is another key factor in successful recruitment and retention programs. Students who are aware of the availability of financial support in science and engineering disciplines are more likely to pursue such careers. Some successful recruitment and retention programs offer forgivable educational loans to students from underrepresented groups who agree to pursue faculty careers (TFMW, 1989).

Although the fundamental problem of encouraging students to pursue careers in science and engineering is not unique to the hydrologic sciences, active pursuit of solutions to the problem is essential to the well-being of the field. At this time, with the increasing need for hydrologic scientists around the world and the expanding opportunities described in this report, the discipline cannot afford to ignore the importance of getting underrepresented groups involved in the hydrologic sciences.

SOURCES AND SUGGESTED READING

Chow, Ven Te. 1959. Open-Channel Hydraulics. McGraw-Hill, New York.

Chow, Ven Te. 1964. Handbook of Applied Hydrology: A Compendium of Water Resources Technology. McGraw-Hill, New York.

Holden, Constance. 1989. Wanted: 675,000 future scientists and engineers. Science 244:1536-1537.

National Research Council (NRC). 1984. Sex Segregation in the Workplace: Trends, Explanations, Remedies. National Academy Press, Washington, D.C.

National Research Council (NRC). 1987. Women: The Underrepresentation and Career Differentials in Science and Engineering. National Academy Press, Washington, D.C.

National Research Council (NRC). 1988. Doctorate Recipients from United States Universities: Summary Report 1987. National Academy Press, Washington, D.C.

National Research Council (NRC). 1989a. Doctorate Recipients from United States Universities: Summary Report 1988. National Academy Press, Washington, D.C.

National Research Council (NRC). 1989b. Responding to the Changing Demography: Women in Science and Engineering. Planning Group to Assess Possible OSEP Initiatives for Increasing the Participation of Women in Scientific and Engineering Careers. Office of Scientific and Engineering Personnel, National Research Council, Washington, D.C.

The Task Force on Women, Minorities, and the Handicapped in Science and Technology (TFWM). 1989. Changing America: The New Face of Science and Engineering. Final Report. National Science Foundation, Washington, D.C.

Vetter, Betty M. 1989. Minorities gain, but white women lose ground. AAAS Observer, September 1, p. 10.

Widnall, S. 1988. Voices from the pipeline. Science 241:1740-1745.

Zikmund, B. 1988. The well-being of academic women is still being sabotaged—by colleagues, by students, and by themselves. Chronicle of Higher Education, September 1, p. A44.

6

Scientific Priorities

Science has developed as a cultural endeavor of value to society for the enrichment of the human spirit that flows from understanding, as well as for the material benefit to which understanding often leads. The culture of science draws its strength from the diversity of interests, experiences, and viewpoints of its individual investigators, and from its traditional principles of integrity and precision.

Few question the wisdom of public financial support for science, but most freely acknowledge practical limits to public altruism. Nations and agencies cannot support every scientific proposal, and thus priorities must be set. Although the political process ultimately determines what gets funded, it is in everyone's best interest that this process have the benefit of knowledgeable scientific opinion. **Scientists must learn to formulate and advocate research priorities in their separate scientific disciplines.**

This chapter outlines a rational process by which priorities might be set to promote a vigorous and beneficial hydrologic science. It includes a few examples of research opportunities that the committee believes to be the most important at this time. In addition it names those developments in education and scientific data viewed as necessary for the vigor of hydrologic science in the long term.

THE PROCESS

The dilemma in the objective establishment of scientific priorities is that the criteria for ranking one opportunity with respect to another are inherently subjective. The 18 members of this committee offered

10 separate but not necessarily independent measures of worth. Each member ranked all 10 in descending order of importance by his or her personal scale of values, and the rankings were averaged across the committee. The interdependence of many of the 10 measures allowed consolidation of the averaged rankings to three criteria:

1. Expected contribution to scientific understanding

Including such measures as either breakthrough or incremental contributions to understanding and reduction of uncertainty, this premier criterion reflects the dominating objective of science.

2. Support of a viable scientific infrastructure

Maintenance of a cadre of hydrologic researchers and synergistic stimulation of related sciences are among the measures leading to this second-ranked criterion.

3. Contribution to problem solving

Social benefits such as the solution of current crises and the optimization of water resources were measures of importance yielding this third-ranked criterion.

THE PREMISES

The diversity and range of scale of the frontier problems in hydrologic science are illustrated clearly by the examples given in Chapter 3. In addressing the issue of priorities among such questions, we must seek to maintain options and diversity, and to keep avenues open for innovation and the operation of serendipity. The above criteria must be augmented with a set of premises:

- **It is not possible to make rational priority judgments among very specific research questions.**

For example, which is more important, the effects of chemistry and biology on soil properties, or how heat and mass flow control water seepage in frozen media? Instead of such fine-grained comparisons, we should consider the relative importance of larger classes of problems, such as land surface-atmosphere interactions versus the generation of streamflow from precipitation.

- **If the number of priority research areas is kept small, the list need not be ranked.**

However, if the categories are too broad they become all-inclusive, and the sense of direction is lost.

- **In selecting the priority areas only the primary criterion should be used.**

Satisfaction of the secondary and tertiary criteria should not be allowed to influence membership on the short list but might be used to rank order this list if desired.

- **The questions with the greatest potential for a contribution to understanding lie at the least-explored scales and in making the linkages across scales.**

Considering that the historical development of hydrologic science began at the small catchment scale and has spread both ways (i.e., both larger and smaller) over time, rich frontiers lie at the global scale and the microscale. Finding the scale-bridging laws of hydrologic similarity will reveal order and pattern.

- **Hydrologic science is currently data-limited.**

Interest in ever-increasing scale has outrun the financial support for observation, and the balance of hydrologic science is now seriously skewed toward modeling. It is important that observation and analysis proceed hand in hand.

PRIORITY CATEGORIES OF SCIENTIFIC OPPORTUNITY (UNRANKED)

In keeping with the above premises, the committee suggests the following five research areas as those now offering the greatest expected contribution to the understanding of hydrologic science.

- **Chemical and Biological Components of the Hydrologic Cycle**

In combination with components of the hydrologic cycle, aqueous geochemistry is the key to understanding many of the pathways of water through soil and rock, for revealing historical states having value in climate research, and for reconstructing the erosional history of continents. Together with the physics of flow in geologic media, aquatic chemistry and microbiology will reveal solute transformations, biogeochemical functioning, and the mechanisms for both contamination and purification of soils and water.

Water is the basis for much ecosystem structure, and many ecosystems are active participants in the hydrologic cycle. Understanding these interactions between ecosystems and the hydrologic cycle is essential to interpreting, forecasting, and even ameliorating global climate change.

• Scaling of Dynamic Behavior

In varied guises throughout hydrologic science we encounter questions concerning the quantitative relationship between the same process occurring at disparate spatial or temporal scales. Most frequently perhaps, these are problems of complex aggregation that are confounding our attempts to quantify predictions of large-scale hydrologic processes. The physics of a nonlinear process is well known under idealized, one-dimensional laboratory conditions, and we wish to quantify the process under the three-dimensional heterogeneity of natural systems, which are orders of magnitude larger in scale. This occurs in estimating the fluxes of moisture and heat across mesoscale land surfaces and in predicting the fluvial transport of a mixture of sediment grains in river valleys. It arises in attempting to extend tracer tests carried out over distances of 10 to 50 m in an aquifer to prediction of solute transport over distances of hundreds of meters to kilometers. It occurs in extrapolating measurements of medium properties in a small number of deep boreholes (as in the Continental Scientific Drilling Program) to characterize fluid fluxes at crustal depth.

The inverse problem, disaggregating conditions at large scale to obtain small-scale information, arises commonly in the parameterization of subgrid-scale processes in climate models and in inferring the subpixel properties of remote sensor images.

Solving these problems will require well-conceived field data collection programs in concert with analysis directed toward "renormalization" of the underlying dynamics. Success will bring to hydrologic science the power of generalization, with its dividends of insight and economy of effort.

• Land Surface-Atmosphere Interactions

Understanding the reciprocal influences between land surface processes and weather and climate is more than an interesting basic research question; it has become especially urgent because of accelerating human-induced changes in land surface characteristics in the United States and globally. The issues are important from the mesoscale upward to continental scales. Our knowledge of the time and space distributions of rainfall, soil moisture, ground water recharge, and evapotranspiration are remarkably inadequate, in part because historical data bases are point measurements from which we have attempted extrapolation to large-scale fields. Our knowledge of their variability, and of the sensitivity of local and regional climates to alterations in land surface properties, is especially poor.

The opportunity now exists for great progress on these issues for the following reasons. Remote sensing tools are available from aircraft and satellites for measurement of many land surface properties. Critical field experiments such as the completed First ISLSCP Field Experiment (FIFE) and the Hydrologic-Atmospheric Pilot Experiments-Modelisation du Bilan Hydrique (HAPEX-MOBILHY), and others under way and planned, promise to improve both measurement and understanding of hydrologic reservoirs and fluxes on several scales. Additional experiments in a range of environments are needed. Finally, numerical models exist that are capable of integrating results from regional and global measurement programs and focusing issues for future experiments.

- **Coordinated Global-scale Observation of Water Reservoirs and the Fluxes of Water and Energy**

Regional- and continental-scale water resources forecasts and many issues of global change depend for their resolution on a detailed understanding of the state and variability of the global water balance. Our current knowledge is spotty in its areal coverage; highly uneven in its quality; limited in character to the quantities of primary historical interest (namely precipitation, streamflow, and surface water reservoirs); and largely unavailable still as homogeneous, coordinated, global data sets. The World Climate Data Program (WCDP) has undertaken the considerable task of assembling the historical and current data, and the World Climate Research Program (WCRP) is planning the necessary global experimental program, the Global Energy and Water Cycle Experiment (GEWEX) (see discussion in Chapter 4), to place future observations on a sound and coordinated scientific foundation. Many nations must contribute for this program to be successful. The United States should play a major role in GEWEX through the support of key experimental components and accompanying modeling efforts. Of particular importance in this regard is NASA's Earth Observing System (EOS) program (see discussion in Chapter 4), which will include observing and data systems as well as scientific experiments for multidisciplinary study of the earth as a system.

- **Hydrologic Effects of Human Activity**

For at least two decades hydrologists have acknowledged that humans are an active and increasingly significant component of the hydrologic cycle. Quantitative forecasts of anthropogenic hydrologic change are hampered, however, by their being largely indistinguishable from the temporal variability of the "natural" system.

Experiment and analysis need to be focused on this question. Identification of the signal of change within the background noise of

SETTING PRIORITIES AMONG SCIENTIFIC INITIATIVES

An inscription on the dome of the Great Hall of the National Academy of Sciences reads: "To science, pilot of industry, conqueror of disease, multiplier of the harvest . . . ." But in this age of tight budgets and expensive science, how does one decide which initiatives to pursue? How does one cope with the reality that science has much more to do than the nation can now afford?

The answer is that judgments must be made. Scientists may argue among themselves in support of their particular disciplines and projects, and refuse to concede any one thing as less important than another, but in the meantime others—most notably politicians—will be making these decisions with or without reasoned assistance. The American scientific community has now recognized this reality and is increasingly involved in seeking ways to set priorities among scientific initiatives. We are seeking consensus on how to evaluate scientific merit, weigh potential benefits to society, and judge the relative feasibility of different proposals.

The goal of science is to produce understanding of physical, chemical, or biological phenomena; thus, scientific endeavors are valued in proportion to the extent that they reveal the laws and interactions governing the structure and evolution of these phenomena. Science needs the stimulus of new discoveries, because without challenges it tends to grow stale, introspective, and concerned with trivia. In general, scientific research includes three stages: (1) exploration and discovery, (2) reconnaissance and observation, and (3) theory and modeling. Different stages may be preeminent in different fields at different times. Any effort to set priorities among initiatives must address the need for balance among these stages as well as requirements related to scientific merit, benefits to society, and relative feasibility.

The necessary task is not to rate competing scientific disciplines but to evaluate scientific initiatives by reasoned means. Formal procedures, using specific criteria, can be developed and consistently used. No evaluation process will satisfy all the parties of interest, but that is to be expected. The objective is to move toward an orderly national science agenda, one that ensures that critical areas are adequately addressed and provides flexibility at the same time as it provides structure. If the scientific community fails to provide broad leadership in formulating such an agenda, the administration and Congress will set their own scientific agenda.

spatial and temporal variability will require observations at regional scale and over many annual cycles. Forecasting the course of future change will be eased by understanding what changes have already occurred.

DATA REQUIREMENTS

• Maintenance of Continuous Long-Term Data Sets

The hydrologic sciences use data that are collected for operational purposes as well as those collected specifically for science. Improvements in the use of operational data require that special attention be given to the maintenance of continuous long-term data sets of established quality and reliability. Experience has shown that exciting scientific and social issues often lead to an erosion in the data collection programs that provide a basis for much of our understanding of hydrologic systems and that document changes in regional and global environments.

• Improved Information Management

The increasing emphasis on global-scale hydrology and the increasing importance of satellite and ground-based remote sensing lead to use of large volumes of data that are collected by many different agencies. An information management system is needed that would allow searching many data bases and integrating data collected at different scales and by different agencies.

• Interpretation of Remote Sensing Data

Effective use of remote sensing data is now too difficult for many hydrologic scientists, because the interpretation often depends on a detailed knowledge of sensor characteristics and electromagnetic properties of the surface and atmosphere. Hydrologic data products should be made available in a form such that scientists who are not remote sensing experts can easily use the information derived.

• Dissemination of Data from Coordinated Experiments

Special integrated studies, such as HAPEX, FIFE, and GEWEX, that involve intensive data collection and investigation of the fluxes of water, energy, sediment, and various chemical species, produce high-quality data sets that have value lasting far beyond the duration of the experiment. Optimal use of these data requires broader and more timely distribution beyond the community of scientists who are involved in the experiments.

EDUCATION REQUIREMENTS

- **Multidisciplinary Graduate Education Program**

The broad range of education inputs to graduate study in hydrologic science necessitates the formation of a multidisciplinary program in the hydrologic sciences. This program should be either a department unit or a confederation of faculty from host departments that is assured of autonomy and resources by upper-level administration. If it is the latter type of organization, it may or may not be degree-granting. The primary purpose of the program would be to educate graduate students who are considered first and foremost as hydrologists, not geologists, geographers, or engineers who have some background in hydrology.

- **Experience with Observation and Experimentation**

The changing nature of hydrologic science requires the development of coordinated, multidisciplinary, large-scale field experiments. Graduate students should be given experience with modern observational equipment and technologies within their university programs, and mechanisms should be developed to facilitate their participation in these experiments, irrespective of their university of study. When the experiments are planned, the inclusion of a diverse array of studies should be an integral part of the plan. Undergraduate students of science should have experience with measurement of natural phenomena, preferably in field situations as well as in controlled laboratory settings.

- **Visibility to Undergraduate Students**

Programs should be developed to make hydrologic science more visible as a scientific discipline to undergraduate students. These programs should include such elements as research participation, internships at laboratories and institutes, curricula that introduce the latest innovations, visiting distinguished lecturers, media development, and in-service institutes for teachers.

SOURCES AND SUGGESTED READING

Dutton, J. A., and L. Crowe. 1988. Setting priorities among scientific initiatives. Am. Sci. 76(Nov.-Dec.):599-603.
Press, F. The Dilemma of the Golden Age. Address to the members of the National Academy of Sciences, April 26, 1988.

Resources and Strategies

The development of a particular science is a cumulative process, with each advance building on the past contributions of diverse fields. Progress is normally the result of steady, patient effort guided and monitored by select disciplinary peer committees, funded by disciplinary programs, and communicated through disciplinary journals. Hydrologic science has not had the benefit of this organized infrastructure, however, as it has grown in response to and been constrained by the evolving engineering and management needs of contemporary societal water problems: first water supply, then flood control, and more recently pollution abatement.

Development of hydrology as a science is vital to the current effort to understand the interactive behavior of the earth system because of the key role that the hydrologic cycle is now known to play therein. Not only is such knowledge prerequisite to solving the many unforeseen water problems that will result from future global change, but it is also needed to cope with the ever-increasing complexity of the more conventional water management problems. Achieving this comprehensive understanding of the earth system will require the kind of long-term disciplinary and interdisciplinary effort that can be sustained only by a vigorous scientific infrastructure. In conclusion this committee presents those resources and strategic actions that it believes are necessary to support a viable hydrologic science in the United States.

RESOURCES

To advance the science of hydrology, resources will be needed in the following areas.

- **Research Grant Programs**

The central role of water in the earth system over a broad range of space and time scales provides the scientific rationale for a unified development of hydrologic science. The associated need to create and maintain a cadre of hydrologic scientists requires development of a focused image and identity for this science. Establishment of distinct but coordinated research grant programs in the hydrologic sciences would address both of these issues.

Support for research in hydrologic science in the United States is scattered among various agencies of the federal government, as detailed in Appendix A. In keeping with the pragmatic origins of the science (summarized in Chapter 2), the "action" agencies, such as the U.S. Geological Survey, the U.S. Environmental Protection Agency (EPA), the National Aeronautics and Space Administration, the National Weather Service, and the Agricultural Research Service, manage water-related research programs oriented to their own specific missions. The basic science fraction of this research, quite properly, is small in comparison with the applied. The amount of funds spent in-house is large with respect to external grants, and there is little coordination of effort at the interagency level.

Support for basic research in hydrologic science is concentrated within the National Science Foundation (NSF) but is diffused there among the divisions of the Geosciences Directorate, each with a mandate oriented toward its own interests. This partitioning not only slights important hydrologic areas, such as aqueous chemistry and the earth's vegetation cover, but also ensures that there is no cultivation of a coherent research program in hydrologic science, and that the science achieves no established identity.

- **Fellowships, Internships, and Instructional Equipment**

The development of education in the hydrologic sciences will require the involvement of scientists, educators, and others in federal, state, or local agencies, who will contribute at all levels—from kindergarten through graduate school.

At the graduate level, this committee recommends establishment of special research fellowships in the hydrologic sciences. These should be designed to train students for research in a specific branch of

COORDINATED FIELD EXPERIMENTS—REGIONAL STUDIES OF LAND-ATMOSPHERE INTERACTION

When the sun evaporates water from the earth's land surface, that process dries and cools that surface. Some of the solar energy is carried with the evaporated water up into the atmosphere, where it is released in turn when that water vapor condenses into rain. In this way the budgets of energy and water are intimately coupled, and the mathematical description of their interrelationship in time and space is one of the principal unsolved problems of hydrologic science. The processes involved depend critically on such things as the physical properties of the soil, the type and density of vegetation cover, the climate, the season of the year, and the weather at a given moment. Clearly, observations are necessary to understand this complex phenomenon, and they must be large enough in spatial scale to obtain useful averages of the earth's naturally heterogeneous surface properties. Observations of both atmosphere and land surface must be made, and these must be coordinated in time and space. Typically, observations of electromagnetic radiation are made from satellites or aircraft; wind, temperature, and atmospheric moisture are measured from sounding balloons; and temperature, soil moisture, and heat and moisture flux are observed at the ground.

The International Land Surface Climatology Project

Part of the World Climate Research Program (WCRP), the International Satellite Land Surface Climatology Project (ISLSCP) has as its objective improving the understanding of interactions between vegetation and the atmosphere on a global scale through the use of observations from satellites. ISLSCP comprises a series of large-scale (20 to 100 km^2) field experiments with the participation of hydrologists, ecologists, atmospheric scientists, and remote sensing experts from the international scientific community.

The First ISLSCP Field Experiment (FIFE) was conducted in the tallgrass prairie region of the United States (near Manhattan, Kansas) in 1987. FIFE consisted of four intensive field campaigns each lasting two to three weeks during June-September, and more than 100 scientists participated in this experiment. The objective of FIFE was to capture the diurnal and seasonal behavior of the tallgrass ecosystem based on a series of in situ, airborne, and space-based conventional and remote sensing observations of the soil-plant-atmosphere system. In 1988 the existing ground-based observation network and satellite sensors were used to monitor the conditions of the tallgrass prairie, without any intensive field campaigns. The preliminary analysis of the comprehensive data sets obtained during the first two years demonstrated a need for further exploration of biophysical control of mass and energy exchange between the prairie ecosystem and the atmosphere during the water-limiting periods. A follow-up intensive field experiment was designed with very specific objectives to be conducted in July-August 1989, coincident with a warm and dry period.

The Hydrologic-Atmospheric Pilot Experiments

The Joint Scientific Committee of the WCRP also initiated the Hydrologic-Atmospheric Pilot Experiments (HAPEX). The first of these experiments, Modelisation du Bilan Hydrique (MOBILHY), was conducted in 1986 southeast of Bordeaux, France (Andre et al., 1986). The focus of HAPEX-MOBILHY was to assess the regional hydrology at the grid scale of numerical climate models (100 × 100 km), based on conventional and remote sensing observations, and with the aid of hydrologic models.

Other Planned Experiments

A number of experiments similar to FIFE and HAPEX are under consideration in a wide variety of ecosystems such as semiarid regions of Spain, grassland regions of West Africa, boreal forest regions between the United States and Canada, steppe regions of the USSR, and the Heiheh River basin in China. FIFE and HAPEX have contributed uniquely to improving our understanding of scale-dependent processes that affect land-atmosphere interactions by bringing together for the first time groups of scientists from diverse disciplines to observe, measure, model, and verify such interactions at regional scales. Follow-on activities planned for the 1990s will enlarge our understanding of scales ranging from the regional to the continental through the proposed Global Energy and Water Cycle Experiment (GEWEX).

The understanding gained from these experiments will be incorporated into the regional and global numerical (i.e., computer-based) models of weather and climate on which society increasingly relies for forecasts of things to come.

hydrology and to increase the number of students equipped to investigate interdisciplinary problems. Travel fellowships will enable students to enroll in specific courses, to interact with key scientists, and to participate in large-scale, coordinated experiments. Fellowships are especially important in increasing participation by women, ethnic minorities, and the handicapped, as are internships for the retraining of mature scientists from allied disciplines.

At the undergraduate level, there is a strong need for providing modern, sensitive instructional equipment for students' use in the field and to back this up with logistical support for field trips and field classes.

Summer or academic year institutes for kindergarten through twelfth grade teachers can provide a basic science and mathematics background taught in the context of hydrology. Under the title of environmental science, or earth science, or general science, interesting hydrologic topics can be developed to fit into everyday instruction in science and mathematics at all levels. A key idea for these institutes is the training of resource teachers who will then conduct workshops in their own schools (or districts) for other teachers. Summer institutes for especially talented science and mathematics students should be established at colleges and universities to stimulate interest in careers in the hydrologic sciences by providing hands-on problem-solving activities.

These activities should be supported primarily through grants from federal agencies—i.e., the EPA, the Department of Energy, the Department of the Interior, the NSF, and others—depending on the connection of the subject matter to their specific missions. However, partial support from state and local governments and industry is both possible and advisable.

• Coordinated Field Experiments

Multidisciplinary field experiments with coordinated observations are needed for answering different types of scientific questions and are useful for instruction in the art and science of field observation. These include short-term, large-scale, multicollaborator studies, sometimes called campaigns or given acronyms such as GEWEX (for Global Energy and Water Experiment); long-term studies of processes, sometimes called base-line studies, such as those of watershed erosion formerly conducted by the Soil Conservation Service; and actual, controlled, small-scale experiments.

Campaigns Simultaneous measurement of many hydrologic processes by scientists working within the context of an agreed-upon plan represents a powerful means of generating new hydrologic insights. This is particularly true when the field program has the participation of theoreticians right from the start of planning and when an intensive effort is made to anticipate how results will be shared and used by various analysts. These investigations (e.g., the Global Atmosphere Research Program/Atlantic Tropical Experiment (GATE) and the Tropical Ocean and Global Atmospheric (TOGA) Program) are widely used by the other geosciences, particularly to characterize mesoscale and larger phenomena, but are just coming into use in hydrology (e.g., HAPEX and FIFE). Even the largest of these field programs can and should be compatible with the best independent and individual science; such a program serves as an umbrella under which individual investigators carry out their work.

Base-Line Studies Some excellent field stations, maintained by federal agencies, now participate in long-term observations in various fields of science. For example, several are part of the national network of the Long-Term Ecological Research (LTER) Program being carried out under the leadership of the NSF's Directorate for Biological, Behavioral, and Social Sciences. Although hydrology is fundamental to many of the research questions studied at the LTER stations, hydrologists are underrepresented among participating scientists; substantial opportunity currently exists for collaboration in ongoing field experiments. There is a current lack of communication among hydrologists in the federal sector and their counterparts in the universities concerning the facilities, resources, and scientific potential of these sites and others. Improved communication is essential to foster the one-on-one contact that is the foundation of developing collaboration. Formalized agency programs supporting faculty and student involvement in field experiments and instruction at these facilities are badly needed.

Observation of Transients Many hydrologic processes are distinguished by extreme episodes of short duration that may be catastrophic in their effect on society (e.g., floods, landslides, hurricanes, and blizzards). Too often these events are investigated weeks or months after their occurrence when the evidence of chronology and mechanism has been degraded. Timely study of these isolated phenomena could be facilitated if funding agencies had a mechanism for releasing funds on extremely short notice to put investigators into the field.

It should be the responsibility of universities and government agencies to inculcate the necessary planning and observational skills for all these modes of research through a steadfast, long-term commitment to the teaching and financial support of field work in the hydrologic sciences.

• Long-Term Observations

Continuous, long-term records of hydrologic-state variables (e.g., soil moisture, temperature, atmospheric humidity, and concentration of dissolved and suspended substances) and hydrologic fluxes (e.g., precipitation, streamflow, and evaporation) are essential, among other things, to quantify the variability of these quantities. Such records can reveal secular trends, periodicities, and the probability distribution of the random residuals—information that has value in such areas, respectively, as identification of global change (the Mauna Loa carbon dioxide record), isolation of mechanisms, and estimation of the risk of flood and drought.

There is no substitute for such long-term data records in science and engineering. Unfortunately, however, their uninterrupted collection and cataloging are an unglamorous task. Therefore, the funds to support this vital task are traditionally high on the budget-cutter's list of targets.

The committee must renew the plea here for unwavering support of the collection and storage of long-term hydrologic records. These resources are like a patient's medical record: useless during apparent health, but invaluable when illness appears. The only certainty is that if records are not kept, they will not be available when needed.

• Access to Data Bases

The immediate, unrefined products of observation and experimentation are scientific data. These are obviously available to those who collect them, but their primary value is often realized by others at a later date and in a quite different scientific context. For hydrologic science to move forward it is essential that data sets, once acquired, be properly identified and described (i.e., purpose, location, instruments, spatial and temporal coverage, and so forth), be cataloged and archived (including archival maintenance), and be made available to the scientific community at reasonable cost and effort. Resources are needed for these tasks.

STRATEGIES

To further the recognition and establishment of hydrologic science as a distinct geoscience, hydrologists can take many actions, either individually or through their scientific societies. These include the following:

• *Make use of relevant scientific societies* as platforms for communication, advocacy, organization, and education. Societies such as the American Geophysical Union, the American Meteorological Society, and the Ecological Society of America can draw attention to the problems and needs of hydrologic science through such activities as preparing, disseminating, and advocating positions on issues of hydrologic science that are of public interest; sponsoring graduate fellowships; generating educational material for use in secondary schools; publishing review articles about hydrologic science directed at readers from allied sciences; and facilitating the organization of international scientific research programs.

LONG-TERM DATA AND THE GREENHOUSE EFFECT

The relationship between certain trace gases in the atmosphere, notably carbon dioxide, and global climate change is described in Chapter 3. Despite the current uncertainty with respect to our understanding of the mechanisms underlying the global hydrologic cycle, a consensus exists that future changes in climate because of a continual increase of trace gas concentrations in the atmosphere may have a significant impact on the survival of ecosystems and the management of water resources.

But how do we know that there has been a continual increase in the concentration of trace gases like carbon dioxide (CO_2) in the atmosphere? In the case of CO_2, we have some indirect information from tree-ring records and from historical measurements of solar absorption spectra, and there is a remarkable set of direct measurements from air bubbles trapped in an ice core taken from Antarctica. But the single most valuable set of CO_2 measurements over time is the record determined by Charles David Keeling at Mauna Loa, Hawaii, beginning in 1958 (Figure 7.1). This very precise time series shows definitively that CO_2 concentrations have been rising from year to year, with a seasonal modulation that also is increasing. It stands as one of the classic examples of long-term field work having major environmental significance.

The Mauna Loa CO_2 measurements were initiated during the International Geophysical Year, but they might have been discontinued shortly thereafter had not a few enlightened people at the Scripps Institution of Oceanography nurtured a young postdoctoral researcher with a mission. Despite the increasingly recognized importance of data records of long duration, only a handful of dedicated research facilities have successfully maintained high-quality data collection sites over long periods. Researchers at these sites have experienced institutional reluctance to commit funds year after year to activity that frequently is termed "monitoring" in a pejorative sense. Emphasis is placed instead on short-term studies expected to bring quick results that can be used to bolster requests for increases in the annual budget.

The problem with this short-term approach is that it is not responsive to the data needs for investigating environmental changes that operate on large spatial scales and long time scales. What would any two- to three-year piece of the graph in the Mauna Loa CO_2 record tell us about greenhouse warming? Recognition must be given to the fact that critical issues of global hydrology cannot be resolved by institutional commitment to support that is limited to the laboratory-research time scale. Commitment and the clarity of vision that inspired that small group at the Scripps Institution three decades ago must now take their place high in the priorities of those who plan and support hydrologic research programs of national scope.

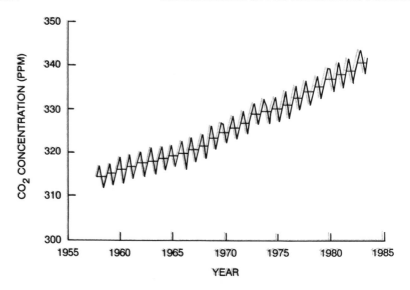

FIGURE 7.1 Concentration of atmospheric carbon dioxide (CO_2) at Mauna Loa Observatory, Hawaii, expressed as a mole fraction in parts per million of dry air. The dots depict monthly averages of visually selected data adjusted to the center of each month. The horizontal bars represent annual averages. SOURCE: Data were obtained by C. D. Keeling, Scripps Institution of Oceanography, University of California, La Jolla, California, and are from files in the Carbon Dioxide Information Center, Oak Ridge National Laboratory, Oak Ridge, Tennessee.

• *Cultivate interest in hydrologic science* among the appropriate mission-oriented agencies of the federal government. There is a need to argue for the allocation of a greater fraction of water research money to be spent by the agencies on basic hydrologic science. There is further need to seek interagency planning and coordination of how these monies are used.

• *Consider the establishment of a separate journal* for hydrologic science. The primary existing outlet for hydrologic science in the United States is *Water Resources Research*, which is arguably the premier journal in the world for a broad spectrum of water research from science to application and policy. It is not widely read in the broader geosciences community, however. On the one hand it is questionable whether the market can yet support a new and separate journal for hydrologic science, but on the other hand the visibility and identity fostered by a separate scientific journal may be necessary to attract the work of scientists from allied fields who want their work to be seen by their peers.

• *Stimulate joint meetings and symposia* among the relevant scientific societies concerning issues of hydrologic science in order to foster interdisciplinary understanding and cooperation.

• *Review, in five years, the progress toward achieving the goals* elaborated in this report, assessing the vitality of the field, surveying the changes that have occurred, and making recommendations for further action.

SOURCES AND SUGGESTED READING

Andre, J. C., J. P. Goutorbe, and A. Perrier. 1986. Hydrologic atmospheric pilot experiment (HAPEX) for the study of water budget and evaporation flux at the atmospheric climate scale. Bull. Am. Meteorol. Soc. 67:138-144.

Becker, F., H. J. Bolle, and P. R. Rowntree. 1988. The International Land-surface Climatology Project. ISLSCP Report No. 10. World Meteorological Organization, Geneva, 100 pp.

World Meteorological Organization. 1988. Concept of the Global Energy and Water Cycle Experiment. WMO/TD–No. 215. WMO, Geneva.

APPENDIXES

Funding for Research in the Hydrologic Sciences

BACKGROUND

In an effort to provide some context for its findings, conclusions, and recommendations, the committee sought to consider and include information in this report on the current (FY 1988), approximate levels of funding for research in the hydrologic sciences. To accomplish this, the staff conducted a survey of the agencies and departments of the U.S. federal government that were expected to have some level of research activity in the hydrologic sciences.

The agencies and departments were given the committee's working definitions of "hydrology"* and "research"† and further information on the scope of the committee's interests. They were asked to provide basic information about their current research programs, including approximate current levels of funding.

Specifically, the committee defined research in the hydrologic sciences as the development of understanding about (1) the paths along

---

*Hydrology—the science concerned with the waters (solid, liquid, and vapor) of the earth; their occurrence, circulation, and distribution; their chemical and physical properties; and their reaction with their environment, including their relation to living things.

This definition was used in the survey, which was begun in the early stages of the committee's work before it had adopted the more comprehensive definition of hydrologic science presented in Chapter 2.

†Research—the development of understanding through the conceptualization and testing of new ideas.

which water moves on and in the land masses of the earth; (2) the rates of movement along various paths; (3) the chemical, biological, and physical character of the constituents transported by water; (4) the controls on the rates of movement of and interactions among water and its transported constituents imposed by the atmosphere, the biosphere, and the oceans; and, conversely (5) the impacts on the atmosphere, geosphere, and biosphere, caused by the presence or absence of water in or on the land. While the committee recognized that scientifically significant information can sometimes be generated in project work, the agencies and departments were asked to exclude from their responses project-related activity that was not of the type that would be reported in the open scientific literature. This information is generally unavailable and unknown to the scientific community.

The agencies were not asked for project details, but instead were asked to attempt to classify their efforts to the extent possible according to the eight contemporary categories of the hydrologic sciences adopted by the committee (e.g., earth crust, climatic processes, living communities, and so on). Further, the agencies were asked to indicate what research was performed in-house by agency staff and otherwise.

SUMMARY OF INFORMATION PROVIDED

Presented in Table A.1 is a summary of agency responses to the survey. Note that the table provides only *total* amounts of research funding levels for the agencies surveyed (with the only breakdowns by agency being the separation of in-house and out-of-house efforts) and the approximate levels of funding by research category across the government.

DISCUSSION

The total amount of funding reported by the federal agencies was approximately $127 million. Of this amount, about $35 million went to sponsor research conducted outside the funding agency. However, not all of the $35 million left the federal sector—some federal agencies conducted research at the behest of their sister agencies, which consumed a small amount of this "out-of-home" funding. A significant amount of that research tabulated as "out-of-home" was also conducted at the national laboratories. Discussions with representatives of the federal agencies indicated that somewhat more than half of the remaining $30 million was dedicated to research that directly supported the operational missions of the funding agencies. Therefore, approximately $10 million to $15 million (about 10 percent) was made available to the nonfederal sector for research that was not necessarily problem driven. For purposes of comparison, in FY 1988 the National Science Foundation provided "unfettered" research funding to the atmospheric- and ocean-science academic communities of $95 million and $135 million, respectively.

To provide additional perspective, the total federal budget for basic research in FY 1990 approximated $10 billion. Therefore, the total amount of funding for research in the hydrologic sciences (about $127 million) is about 1 percent of the total federal expenditures for basic research.

Although the committee believes that the majority of hydrologic research is either conducted or sponsored by the federal agencies surveyed, additional research in the hydrologic sciences may be carried out under the aegis of states, industries, and private foundations. A more exhaustive search may have produced a slightly larger total.

Discussions with the federal representatives also indicated that they had difficulty in accurately partitioning the research funds that they managed into the eight categories of the hydrologic sciences adopted by the committee. Therefore Table A.1 does not report funding by agency for the individual categories, but instead indicates with a symbol (•) that an agency conducts or funds hydrologic science in a given category. In spite of the lack of accuracy of the individually reported numbers, it is interesting to note that the two categories receiving the largest amounts of funding were "Earth Crust" and "Chemical Processes." These categories encompass about one-half of the total, which is not surprising in light of the current federal interest in ground water contamination.

TABLE A.1 Federal Funding for Research in the Hydrologic Sciences (figures in $ million for Fiscal Year 1988)

	National Aeronautics and Space Administration		U.S. Geological Survey		U.S. Fish and Wildlife Service		U.S. Bureau of Reclamation		National Science Foundation		U.S. Environmental Protection Agency		Army Corps of Engineers	
	I	O	I	O	I	O	I	O	I	O	I	O	I	O
TOTAL FUNDING	1.2	2.3	48.3	4.2	0.2		0.2			6.3	7.9	5.8	0.8	
1 Earth Crust			•	•			•			•	•	•	•	
2 Land Forms		•	•	•						•			•	
3 Climatic Processes	•	•	•	•						•			•	
4 Weather Processes	•		•	•						•			•	
5 Surficial Processes	•	•	•	•			•			•	•	•	•	
6 Living Communities	•	•	•	•	•					•	•	•	•	
7 Chemical Processes			•	•			•			•	•	•	•	
8 Additional Topics										•				

NOTE: I = In-house
O = Extramural

Army Research Office		U.S. Nuclear Regulatory Commission		Tennessee Valley Authority		U.S. Department of Energy		Forest Service		Agricultural Research Service		National Oceanic and Atmospheric Administration (including National Weather Service)		Total Approximate Federal Level of Funding for Research Categories		
I	O	I	O	I	O	I	O	I	O	I	O	I	O	I	O	TOTAL
	0.7		1.9	1.1			8.7	9.5	5.0	21.1			1.1	92.1	34.9	127.0
			•				•							16.0	6.7	22.7
	•							•		•	•			15.2	2.3	17.5
										•			•	4.8	1.4	6.2
	•							•		•		•	•	6.0	1.7	7.7
								•		•			•	4.6	2.8	7.4
							•	•		•	•			14.8	10.4	25.2
	•		•	•			•	•		•	•		•	29.2	9.1	38.3
										•				1.5	0.5	2.0

Profiles of the Hydrologic Community, 1960 and 1988

This appendix reveals a changing profile of the hydrologic community by contrasting data obtained in a recent survey as part of this study with information published nearly three decades ago.

The 1960 data, taken from "Scientific Hydrology" (Federal Council for Science and Technology, 1962), were obtained from the National Register of Scientific Personnel, which was discontinued in 1971. The Register in 1960 included 811 individuals who called themselves hydrologists.

The data for 1988 were obtained by surveying the approximately 3,000 members of the American Geophysical Union's Hydrology Section. A survey form (Figure B.1, which follows Table B.4) was sent to each member; about 2,200 individuals responded. Of these respondents about 50 did not follow instructions to the extent that their returns were rendered unusable.

Table B.1 compares personal data on hydrologists; Table B.2, their educational backgrounds; Table B.3, employment of hydrologists; and Table B.4, their areas of specialization.

Comment on Possible Bias

The 1960 sample was presumably drawn from across the full spectrum of hydrologists employed in science and technology at that time. The 1988 sample was drawn solely from the membership of a scientific society, the American Geophysical Union. A sampling bias is therefore possible that might lead to underrepresentation, in the 1988 survey, of hydrologists in the consulting engineering sector, of those with an engineering educational background, and of those with less than a doctoral degree.

TABLE B.1 Personal Data on Hydrologists

	Respondents 1960 (percent)	Respondents 1988		
		All (percent)	Male (number)	Female (number)
Gender				
Male	N/A	89		
Female	N/A	11		
Age distribution				
20-24	2	<1	2	0
25-29	7	8	119	44
30-34	15	20	327	95
35-39	18	22	421	55
40-44	13	13	271	17
45-49	14	11	234	6
50-54	16	9	178	6
55-69	13	12	271	3
70 and over	2	4	99	2

TABLE B.2 Educational Backgrounds of Hydrologists

	Respondents 1960 (percent)	Respondents 1988		
		All (percent)	Male (number)	Female (number)
Level of education				
Less than a bachelor's degree	4	<1	6	1
Bachelor's degree	74	11	212	29
Master's degree	17	36	639	133
Professional degree	N/A	1	26	0
Doctoral degree	5	51	1,039	65
Field of highest degree				
Agriculture	N/A	1	25	0
Engineering	55	35	698	49
Environmental sciences	N/A	5	73	21
Forestry	N/A	2	36	1
Mathematics, physics, chemistry	6	4	73	7
Hydrology	N/A	14	276	29
Geography	N/A	3	55	12
Geology	28	28	529	82
Soil science	N/A	3	67	4
Meteorology	6	N/A	N/A	N/A
Other	5	5	90	23

TABLE B.3 Employment of Hydrologists

| | Respondents 1960 (percent) | Respondents 1988 | | |
		All (percent)	Surface Water (number)	Ground Water (number)
Location of employment				
Educational institutions	6	27	277	240
State and local government	11	7	59	81
Industry, business, and self-employed	10	32	172	515
Federal government	65	30	289	276
Military	1	<1	2	3
Nonprofit organizations	1	1	8	22
Others	6	2	18	27
Work activity distribution				
Teaching and university research	N/A	26		
Other research	N/A	21		
Consulting, engineering and other applied hydrology	N/A	40		
Management, administration and regulation	28	13		
Research, development, or design	30	N/A		
Teaching	4	N/A		
Production and inspection	5	N/A		
Other, including no report	33	N/A		

TABLE B.4 Hydrologic Specialties

	Respondents 1960 (percent)	Respondents 1988			
		All (percent)	B.S. (no.)	M.S. (no.)	Ph.D. (no.)
Traditional principal specialties					
Surface water	63	38			
Ground water	27	54			
Snow, ice, permafrost	2	2			
Glaciology	2	2			
Other	6	4			
Contemporary scientific specialties					
Earth crust	N/A	26	70	247	239
Land forms	N/A	12	20	70	156
Climatic processes	N/A	4	10	14	57
Weather processes	N/A	6	29	36	73
Surficial processes	N/A	13	28	77	156
Living communities	N/A	2	2	10	28
Chemical processes	N/A	23	54	225	204
Data technologies	N/A	6	19	44	66
Other	N/A	8	9	49	125

MEMORANDUM TO: Members, AGU Section of Hydrology

From: P.S. Eagleson, Chairman
               NRC Committee on Opportunities in the Hydrologic Sciences

Date: September 26, 1988

Subject: Profile of Hydrologic Community

    The National Research Council's Committee on Opportunities in the Hydrologic Sciences needs to compare the profile (i.e., education background and level, specialty, employer, etc.) of those calling themselves hydrologists today with that of a similar sample done in 1960.* This should be an indicator of educational and employment trends and as such will be helpful to the work of our Committee and should be of interest to you.

    Please take the two minutes needed to complete the enclosed questionnaire and return it in the enclosed envelope. The results will appear in EOS. Thank you.

*Federal Council for Science and Technology, "Scientific Hydrology", June 1962.

QUESTIONNAIRE TO PRIMARY AFFILIATES, AGU SECTION OF HYDROLOGY

A. Level of Education
    __ Less than a bachelor's degree
    __ Bachelor's degree
    __ Master's degree
    __ Professional degree
    __ Doctoral degree

B. Field of Highest Degree (check one)
    __ Agriculture   __ Hydrology
    __ Engineering   __ Geography
    __ Environmental Science   __ Geology
    __ Forestry   __ Soil Science
    __ Mathematics, Physics,   __ Other
        Chemistry

C. Employment (check one)
    __ Educational institutions   __ Federal Government
    __ State and local government   __ Military
    __ Industry, business and   __ Nonprofit organizations
        self-employed   __ Others

D. Work Activity (check one)
    __ Teaching and university research
    __ Other research
    __ Consulting engineering and other applied hydrology
    __ Management, administration, regulation

E. Hydrologic Specialty (check one "Traditional" category <u>and</u> one
    "Contemporary Scientific" category)

<u>Traditional</u>

__ Surface water   __ Glaciology
__ Ground water   __ Other
__ Snow, ice, permafrost

<u>Contemporary Scientific</u>

__ Earth Crust (i.e., ground water and associated heat and mass
        transfer, etc.)
__ Land Forms (i.e., erosion, deposition, and fluvial
        geomorphology, etc.)
__ Climatic Processes (i.e., global water balance, interaction of
        land surface and climate, paleohydrology, etc.)
__ Weather Processes (i.e., space-time precipitation, flash floods,
        interaction of land surface and mesoscale weather systems,
        etc.)
__ Surficial Processes (i.e., infiltration, evaporation, snowmelt,
        etc.)
__ Living Communities (i.e., relationships between vegetation
        patterns and climate, metabolism and energetics of microbial
        communities in water, etc.)
__ Chemical Processes (i.e., geochemical characterization of surface
        and ground waters, etc.)
__ Data Technologies (i.e., remote sensing, computer systems, etc.)
__ Other (i.e., applied mathematics for hydrology such as fractals,
        chaos, etc.)

F. Age Distribution
    __ 20-24   __ 45-49
    __ 25-29   __ 50-54
    __ 30-34   __ 55-69
    __ 35-39   __ 70 and over
    __ 40-44

G. Gender
    __ Male   __ Female

Contributors to the Report, *Opportunities in the Hydrologic Sciences*

During the course of this study, numerous persons other than those listed in the front matter played roles in the development of this report. Some reviewed the work of the committee, others provided material or advice at the committee's invitation, and still others provided unsolicited material for the committee's consideration. The committee carefully considered all of this assistance and wishes to acknowledge and thank these persons for their interest and cooperation. The committee notes that not all material received was published and that not all suggestions were heeded. We acknowledge the possibility of overlooking some individual from our listing and apologize if this occurred.

WILLIAM ALLEY, U.S. Geological Survey, Reston
MARY P. ANDERSON, University of Wisconsin, Madison
LELANI L. ARRIS, Seattle, Washington
VICTOR R. BAKER, University of Arizona
ROGER BALES, University of Arizona
ERIC J. BARRON, Pennsylvania State University
THOMAS L. BELL, NASA Goddard Space Flight Center
KENNETH BENCALA, U.S. Geological Survey
KEITH BEVEN, University of Lancaster, United Kingdom
ISTVAN BOGARDI, University of Nebraska-Lincoln
JOHN J. BOLAND, The Johns Hopkins University
RAFAEL L. BRAS, Massachusetts Institute of Technology
JOHN D. BREDEHOEFT, U.S. Geological Survey, Menlo Park

CAROL BREED, U.S. Geological Survey
NATHAN BURAS, University of Arizona
STANLEY BUTLER, University of Southern California
MICHAEL E. CAMPANA, Desert Research Institute, University of
   Nevada-Reno
MOUSTAFA CHAHINE, California Institute of Technology,
   Jet Propulsion Laboratory
DAVID A. DAUGHARTY, University of New Brunswick
PAUL R. DAY, University of California-Berkeley
ROBERT E. DICKINSON, National Center for Atmospheric
   Research, Boulder, Colorado
WILLIAM E. DIETRICH, University of California-Berkeley
JAMES C. I. DOOGE, Dublin, Ireland
JOHN DRACUP, University of California-Los Angeles
CHARLES T. DRISCOLL, Syracuse University
LUCIEN DUCKSTEIN, Case Western Reserve University
CHRIS J. DUFFY, Utah State University
DARA ENTEKHABI, University of Arizona
MALIN FALKENMARK, NFR, Stockholm, Sweden
R. ALLAN FREEZE, University of British Columbia
JURGEN GARBRECHT, USDA, Agricultural Research Service
WILFORD R. GARDNER, University of California-Berkeley
KONSTANTINE P. GEORGAKAKOS, University of Iowa
EDWARD M. GODSY, U.S. Geological Survey, Menlo Park
DAVID C. GOODRICH, USDA, Agricultural Research Service
VERNON HAGEN, Dewberry and Davis Engineers
HAROLD HEMOND, Massachusetts Institute of Technology
WILLIAM H. HENDERSHOT, McGill University
JANET G. HERING, Swiss Institute for Water Resources and Water
   Pollution Control
ROBERT M. HIRSCH, U.S. Geological Survey, Reston
MICHAEL HUDLOW, NOAA, National Weather Service
BRYAN L. ISACKS, Cornell University
RAY D. JACKSON, USDA, Agricultural Research Service, Phoenix
DEAN S. JEFFRIES, National Water Research Institute, Canada
DOUGLAS L. KANE, University of Alaska-Fairbanks
RICHARD C. KATTELMANN, University of California-Santa Barbara
VANCE C. KENNEDY, U.S. Geological Survey, Menlo Park
DAVID F. KIBLER, Pennsylvania State University
M. J. KIRKBY, University of Leeds, United Kingdom
VIT KLEMEŠ, International Association of Hydrological Sciences
LEONARD F. KONIKOW, U.S. Geological Survey, Reston
M. KUHN, Innsbruck, Austria

R. G. LAWFORD, Environment Canada
SHAUN LOVEJOY, McGill University
DAVID R. MAIDMENT, University of Texas
NICHOLAS MATALAS, U.S. Geological Survey, Reston
JOHN R. MATHER, New Jersey
A. I. MCKERCHAR, New Zealand Department of Scientific and
    Industrial Research
MARK F. MEIER, University of Colorado-Boulder
HAROLD A. MOONEY, Stanford University
FRANCOIS M. M. MOREL, Massachusetts Institute of Technology
CATHERINE NICOLIS, Institute d'Aeronomie Spatiale de Belgique
BRIGID O'FARRELL, National Research Council, Commission on
    Behavioral and Social Sciences and Education
AMY PARKER, Annandale, Virginia
EUGENE PATTEN, U.S. Geological Survey, Reston
GEORGE F. PINDER, University of Vermont
NEIL PLUMMER, U.S. Geological Survey, Reston
KENNETH POTTER, University of Wisconsin-Madison
RONALD PRINN, Massachusetts Institute of Technology
EUGENE RASMUSSEN, University of Maryland
JOHN SCHAAKE, NOAA, National Weather Service
WILLIAM G. SHOPE, JR., U.S. Geological Survey, Reston
S. E. SILLIMAN, University of Notre Dame
KEITH V. SLACK, U.S. Geological Survey, Menlo Park
RICHARD A. SMITH, U.S. Geological Survey, Reston
A. F. SPILHAUS, JR., American Geophysical Union
DONALD E. STREBEL, NASA Goddard Space Flight Center
A. VAN DER BEKEN, Vrije Universiteit, Brussels
T. S. UITERKAMP, Netherlands Organization for Applied Scientific
    Research
JACK B. WAIDE, U.S. Forest Service, Oxford, Mississippi
RICHARD H. WARING, Oregon State University
EDWARD WAYMIRE, Cornell University
JACKSON R. WEBSTER, Virginia Polytechnic Institute and State
    University
WILLIAM W. WOESSNER, University of Montana
M. GORDON WOLMAN, The Johns Hopkins University
WALTER O. WUNDERLICH, Knoxville, Tennessee
ISZTAR ZAWADSKI, University of Quebec, Canada
F. C. ZUIDEMA, National Council for Agricultural Research,
    The Netherlands

Biographical Sketches of Committee Members

PETER S. EAGLESON (Chairman) received his B.S. and M.S. in civil engineering from Lehigh University in 1949 and 1952, respectively. He received his Sc.D. from Massachusetts Institute of Technology in 1956 and is currently professor of civil engineering there. Originally a fluid dynamicist, he moved into hydrology in 1964. Recently his research interest has been hydroclimatology, with particular attention to questions of global scale. Dr. Eagleson is a member of the National Academy of Engineering and of the National Research Council's Commission on Geosciences, Environment, and Resources, and he was a founding member of the Water Science and Technology Board. He is a past president of the American Geophysical Union.

WILFRIED H. BRUTSAERT received his B.S. from the State University of Ghent in Belgium in 1958, and his M.Sc. and Ph.D. in engineering from the University of California in 1960 and 1962, respectively. His area of expertise is hydrology. Presently he is professor of hydrology at Cornell University. His areas of research include flow through porous media, permeability, infiltration and drainage, microclimatology, evaporation, surface water hydrology, and hydrologic systems.

SAMUEL C. COLBECK obtained his B.S. and M.S. from the University of Pittsburgh in 1962 and 1965, respectively. He received his Ph.D. in geophysics from the University of Washington in 1970. His area of expertise is properties of snow and ice. Currently he is a geophysicist with the U.S. Army Cold Regions Research and Engineering Laboratory. Previously he was adjunct professor of engineering at Dartmouth College. His areas of research include the properties of snow, including mechanics, electromagnetics, metamorphism, water, and heat flow; and growth of ice crystals from both melt and vapor.

KENNETH W. CUMMINS acquired his B.A. from Lawrence College in 1955 and his M.S. in fisheries and wildlife and Ph.D. in zoology, both from the University of Michigan, in 1957 and 1961, respectively. His area of expertise is aquatic ecology. Previously he held positions at Michigan State University, Oregon State University, and the University of Maryland. He is currently director of the Pymatuning Laboratory of Ecology of the University of Pittsburgh. His research interests include the structure and function of stream ecosystems.

JEFF DOZIER obtained his B.A. in geography from California State University, Hayward in 1968 and his M.Sc. and Ph.D. in geography from the University of Michigan in 1969 and 1973, respectively. He currently holds a joint appointment as professor of geography at the University of California, Santa Barbara, where he has taught since 1974, and as a senior member of the technical staff of the Jet Propulsion Laboratory, California Institute of Technology. His research interests include snow hydrology, remote sensing, terrain analysis, and image processing. He is also the chairman of the Polar Research Board's Committee on Glaciology of NASA's Science Advisory Panel for the EOS Data and Information System.

THOMAS DUNNE is a hydrologist and geomorphologist who is a professor of geology at the University of Washington. He holds a Ph.D. in geography from The Johns Hopkins University. His research interests include hillslope hydrology and geomorphology, and fluvial geomorphology and sedimentation. He is also a member of the Water Science and Technology Board's Committee on Water Resources Research and was elected to the National Academy of Sciences in 1988.

JOHN M. EDMOND acquired his B.Sc. in chemistry from the University of Glasgow in 1965 and his Ph.D. in marine chemistry from the University of California, San Diego in 1970. Currently he is professor of marine chemistry at Massachusetts Institute of Technology. His research interests include processes and mechanisms controlling the composition of oceanic and continental waters and sediments in space and time. He is a member of the National Research Council's Ocean Studies Board.

VIJAY K. GUPTA received his M.S. from Colorado State University and his Ph.D. in hydrology from the University of Arizona. Currently he is professor of hydrology at the University of Colorado-Boulder. Until recently, he was professor of civil engineering at the University of Mississippi and adjunct professor of civil engineering at Utah State University. His research has focused on scientific aspects of water and solute transport in porous media, statistical models of space-time rainfall, analytical modeling of hydrologic processes at the basin scale, and statistical structure of streamflows, including extremes.

GORDON C. JACOBY obtained his Ph.D. in hydrogeology from Columbia University in 1971. Currently he is a research scientist with the Tree-Ring Laboratory at the Lamont-Doherty Geological Observatory of Columbia University. He is also adjunct associate professor with the Department of Geology and Geography at the University of Massachusetts, Amherst.

SYUKURO MANABE acquired his B.S., M.A., and D.Sc. from Tokyo University in 1953, 1955, and 1958, respectively. His areas of expertise are climate dynamics and climate modeling. Currently he is a research meteorologist of climate modeling at the Geophysics Fluid Dynamics Laboratory of the National Oceanic and Atmospheric Administration at Princeton University. His areas of research include physical mechanisms for climate variation by use of mathematical models of climate and climate change resulting from the future increase of atmospheric carbon dioxide. Dr. Manabe was elected to membership in the National Academy of Sciences in 1990, just prior to the completion of this project.

SHARON E. NICHOLSON received her B.S., M.S., and Ph.D. in meteorology from the University of Wisconsin in 1971, 1972, and 1976, respectively. Currently she is associate professor of meteorology at Florida State University. Previously she was adjunct assistant professor of physics at Clark University, assistant professor in the Graduate School of Geography at Clark University, and research associate in the Department of Environmental Sciences at the University of Virginia. Her areas of interest include tropical meteorology, climatic change, paleoclimatology and historical climatology, drought and arid lands, urban climatology and microclimatology, and remote sensing.

DONALD R. NIELSEN is professor of soil science at the University of California-Davis and chairman of the Department of Agronomy and Range Science. He holds a Ph.D. in soil physics from Iowa State University and specializes in research topics such as monitoring water and solute movement within and below the root zone of plants and assessing the spatial variability of water-conducting properties of field soils.

IGNACIO RODRIGUEZ-ITURBE received his M.S. from California Institute of Technology in 1965 and his Ph.D. from Colorado State University in 1967. Currently he is professor of civil and environmental engineering at the University of Iowa. Until recently, he was professor at the Instituto Internacional de Estudios Avanzados in Caracas. His academic and professional interests include analysis, synthesis, and sampling of hydrologic processes; stochastic modeling of natural phenomena; and design of data collection networks. Dr. Rodriguez-Iturbe is a foreign associate of the National Academy of Engineering.

JACOB RUBIN obtained his Ph.D. in soil physics from the University of California-Berkeley. Currently he is a research soil scientist with the National Research Program of the U.S. Geological Survey's Water Resources Division, Menlo Park, California. He conducts and supervises research on water flow in unsaturated porous media and on transport of reacting solutes in sediments and soils. He is also a consulting professor at Stanford University, where he has taught and also supervised Ph.D. students.

J. LESLIE SMITH acquired his B.S. at the University of Alberta in 1974 and his Ph.D. at the University of British Columbia in 1978. Currently he is a professor with the Department of Geological Sciences at the University of British Columbia. His areas of research include stochastic simulation of fluid flow and solute transport in porous media, transport processes in fractured rocks, and the role of ground water flow in geologic and geodynamic processes.

GARRISON SPOSITO received his B.S. and M.S. in soil science from the University of Arizona in 1961 and 1963, respectively. He received his Ph.D. in soil science from the University of California, Berkeley in 1965. He is currently professor of Soil Physical Chemistry at Berkeley. Previously he served on the faculties of the University of California, Riverside and Sonoma State University. His areas of research include the surface chemistry of soils, metal-organic matter reactions, computer simulation of soil solutions, and the physics of mass transport in porous media.

WAYNE T. SWANK obtained his B.S. in forestry from West Virginia University and his M.F. and Ph.D. from the University of Washington. His area of expertise is forest hydrology and ecology. At present he is project leader at the Coweeta Hydrologic Laboratory, USDA, Forest Service and adjunct professor in botany at the University of Georgia. Previously, he served as program director of ecosystem studies at the National Science Foundation. His areas of research include hydrologic process studies in the context of watershed-scale experiments, nutrient cycling and forest productivity, and atmospheric deposition to vegetated surfaces.

EDWARD J. ZIPSER received his B.S.E. in aeronautical engineering from Princeton University in 1958 and his M.S. and Ph.D. in meteorology from Florida State University in 1960 and 1965, respectively. He is head of the Department of Meteorology at Texas A & M University. Previously he was a senior scientist at the National Center for Atmospheric Research, including a period as director of the Convective Storms Division. He has organized national and international field programs to investigate the properties of storm systems. His research expertise is in mesoscale and tropical meteorology.

STEPHEN BURGES (ex officio) acquired his B.Sc. in physics and mathematics and B.E. in civil engineering at the University of Newcastle, Australia, in 1967. He received an M.S. (1968) and Ph.D. (1970) in civil engineering from Stanford University. He has been a member of the faculty at the University of Washington since 1970 and currently is a professor of civil engineering. Dr. Burges was a member of the Water Science and Technology Board until July 1989.

Index